D1187650

Misconceptions of Risk

Misconceptions of Risk

Terje Aven

University of Stavanger, Norway

A John Wiley and Sons, Ltd., Publication

This edition first published 2010

Registered office
John Wiley & Sons Ltd, The Atrium, Southern Gate, Chichester, West Sussex, PO19 8SQ, United Kingdom

For details of our global editorial offices, for customer services and for information about how to apply for permission to reuse the copyright material in this book please see our website at www.wiley.com.

Library of Congress Cataloging-in-Publication Data

Record on file

A catalogue record for this book is available from the British Library

ISBN: 978-0-470-68388-0 (HB)

Typeset in 10/12pt Times by Laserwords Private Limited, Chennai, India
Printed and bound in Great Britain by TJ International, Padstow, Cornwall.

Contents

Preface

We all face risks of some sort, as individuals, businesses and society. We need to understand, describe, analyse, manage and communicate these risks, and the discipline of risk assessment and risk management has been developed to meet this need. This discipline is growing rapidly and there is an enormous drive and enthusiasm to implement risk assessment methods and risk management in organizations. There are great expectations that these tools provide suitable frameworks for obtaining high levels of performance and balancing different concerns such as safety and costs. But the analysis and management of risk are not straightforward. There are many challenges. The risk discipline is young and there are a number of ideas and conceptions of risk out there. For example, many analysts and researchers consider it appropriate to base their risk management policies on the use of expected values, which basically means that potential losses are multiplied with their associated consequences. However, the rationale for such a policy is questionable when facing situations with large uncertainties – an expected value could produce poor predictions of the actual outcome. Another example is the conception that a risk characterization can be based on probabilities alone. However, a probability assignment or a probability estimate is always conditional on background knowledge and surprises could occur relative to these assessments. Hence, risk extends not only beyond expected values, but also beyond probabilities.

A number of such common conceptions of risk have been identified, altogether 19 in number. These conceptions are formulated as headings of the following 19 chapters. The conceptions are discussed and through argumentation and examples their support, strengths, weaknesses and limitations are revealed. The conclusion is that they are often better judged as *mis*conceptions of risk than conceptions of risk. The final chapter provides my overall conclusions on the issues addressed in the book based on the discussions set out in the previous chapters.

The book has been written for professionals in the risk field, including researchers and graduate students. All those working with risk-related problems need to understand the fundamental ideas and concepts of risk. The book is (conceptually) advanced but at the same time easy to read. It has been a goal to provide a simple analysis without compromising on the requirement for precision and accuracy. Technicalities are reduced to a minimum, while ideas and principles are highlighted. Reading the book requires no special background, but for certain parts a basic knowledge of probability theory and statistics is required. It has, however, been a goal to reduce the dependence on extensive prior knowledge of probability theory and statistics. The key statistical concepts will be introduced and discussed thoroughly in the book. Boxes are used to indicate material that some readers would find technical.

The book is about fundamental issues in risk analysis and risk management, and it provides recommendations and guidance in this context. It is, however, not a recipe book, and does not tell you which risk analysis methods should be used in different situations. What is covered is the general thinking process related to the understanding of risk, and how we should describe, analyse, evaluate, manage and communicate risk. Examples are provided to illustrate the ideas.

Acknowledgements

Many people have provided helpful comments on and suggestions for this book. In particular, I would like to acknowledge Eirik B. Abrahamsen and Roger Flage for the great deal of time and effort they spent on reading and preparing comments on earlier versions of the book. I am also grateful to an anonymous reviewer for valuable comments and suggestions.

For financial support, thanks to the University of Stavanger and the Research Council of Norway.

I also acknowledge the editing and production staff at John Wiley & Sons for their careful and effective work.

1

Risk is equal to the expected value

If you throw a die, the outcome will be either 1, 2, 3, 4, 5 or 6. Before you throw the die, the outcome is unknown – to use the terminology of statisticians, it is random. You are not able to specify the outcome, but you are able to express how likely it is that the outcome is 1, 2, 3, 4, 5 or 6. Since the number of possible outcomes is 6 and they are equally probable – the die is fair – the probability that the outcome turns out to be 3 (say), is 1/6. This is simple probability theory, which I hope you are familiar with.

Now suppose that you throw this die 600 times. What would then be the average outcome? If you do this experiment, you will obtain an average about 3.5. We can also deduce this number by some simple arguments: about 100 throws would give an outcome equal to 1, and this gives a total sum of outcomes equal to 100. Also about 100 throws would give an outcome equal to 2, and this would give a sum equal to 2 times 100, and so on. The average outcome would thus be

$$(1 \times 100 + 2 \times 100 + 3 \times 100 + 4 \times 100 \\ + 5 \times 100 + 6 \times 100)/600 = 3.5. \quad (1.1)$$

In probability theory this number is referred to as the expected value. It is obtained by multiplying each possible outcome with the associated probability, and summing over all possible outcomes. In our example this gives

$$1 \times 1/6 + 2 \times 1/6 + 3 \times 1/6 + 4 \times 1/6 + 5 \times 1/6 + 6 \times 1/6 = 3.5. \quad (1.2)$$

We see that formula (1.2) is just a reformulation of (1.1) obtained by dividing 100 by 600 in each sum term of (1.1). Thus the expected value can be interpreted as the average value of the outcome of the experiment if the experiment is repeated

Misconceptions of Risk T. Aven

over and over again. Statisticians would refer to the law of large numbers, which says that the average value converges to the expected value when the number of experiments goes to infinity.

Reflection

For the die example, show that the expected number of throws showing an outcome equal to 2 is 100 when throwing the die 600 times.

In each throw, there are two outcomes: one if the outcome is a 'success' (that is, shows 2), and zero if the outcome is a 'failure' (that is, does not show 2). The corresponding probabilities are 1/6 and 5/6. Hence the expected value for a throw equals $1 \times 1/6 + 0 \times 5/6 = 1/6$, in other words the expected value equals the probability of a success. If you perform 2 throws the expected number of successes equals $2 \times 1/6$, and if you perform 600 throws the expected number of successes equals $600 \times 1/6 = 100$. These conclusions are intuitively correct and are based on a result from probability calculus saying that the expected value of a sum equals the sum of the expected values. Thus the desired result is shown. □

The expected value is a key concept in risk analysis and risk management. It is common to express risk by expected values. Here are some examples:

- For some experts 'risk' equals expected loss of life expectancy (HM Treasury, 2005, p. 33).
- Traditionally, hazmat transport risk is defined as the expected undesirable consequence of the shipment, that is, the probability of a release incident multiplied by its consequence (Verma and Verter, 2007).
- Risk is defined as the expected loss to a given element or a set of elements resulting from the occurrence of a natural phenomenon of a given magnitude (Lirer *et al.*, 2001).
- Risk refers to the expected loss associated with an event. It is measured by combining the magnitudes and probabilities of all of the possible negative consequences of the event (Mandel, 2007).
- Terrorism risk refers to the expected consequences of an existent threat, which, for a given target, attack mode, target vulnerability and damage type, can be expressed as the probability that an attack occurs multiplied by the expected damage, given that an attack occurs (Willis, 2007).
- Flood risk is defined as expected flood damage for a given time period (Floodcite, 2006).

But is an expected value an adequate expression of risk? And should decisions involving risk be based on expected values?

Example. A Russian roulette type of game

Let us look at an example: a Russian roulette type of game where you are offered a play using a six-chambered revolver. A single round is placed in the revolver such that the location of the round is unknown. You take the weapon and shoot, and if it discharges, you lose $24 million. If it does not discharge, you win $6 million.

As the probability of losing $24 million is 1/6, and of winning $6 million is 5/6, the expected gain is given by

$$-24 \times 1/6 + 6 \times 5/6 = -4 + 5 = 1.$$

Thus the expected gain is $1 million. Say that you are not informed about the details of the game, just that the expected value equals $1 million. Would that be sufficient for you to make a decision whether to play or not play? Certainly not – you need to look beyond the expected value. The possible outcomes of the game and the associated probabilities are required to provide the basis for an informed decision. Would it not be more natural to refer to this information as risk, and in particular the probability that you lose $24 million? As we will see in coming chapters, such conceptions of risk are common.

The game has an expected value of $1 million, but that does not mean that you would accept the game as you may lose $24 million. The probability 1/6 of losing may be considered very high as such a loss could have dramatic consequences for you. And how important is it for you to win the $6 million? Perhaps your financial situation is good and an additional $6 million would not change your life very much for the better. The decision to accept the play needs to take into account aspects such as usefulness, desirability and satisfaction. Decision analysts and economists use the term utility to convey these aspects.

Daniel Bernoulli: The need to look beyond expected values

The observation that there is a need for seeing beyond the expected values in such decision-making situations goes back to Daniel Bernoulli (1700–1782) more than 250 years ago. In 1738, the *Papers of the Imperial Academy of Sciences in St Petersburg* carried an essay with this central theme: 'the value of an item must not be based on its price, but rather on the utility that it yields' (Bernstein, 1996). The author was Daniel Bernoulli, a Swiss mathematician who was then 38 years old. Bernoulli's St Petersburg paper begins with a paragraph that sets forth the thesis that he aims to attack (Bernstein, 1996):

> Ever since mathematicians first began to study the measurement of risk, there has been general agreement on the following proposition: Expected values are computed by multiplying each possible gain by the number of ways it can occur, and dividing the sum of these products by the total number of cases.

Bernoulli finds this thesis flawed as a description of how people in real life go about making decisions, because it focuses only on gains (prices) and probabilities, and not the utility of the gain. Usefulness and satisfaction need to be taken into account. According to Bernoulli, rational decision-makers will try to maximize expected utility, rather than expected values (see Chapter 6). The attitude to risk and uncertainties varies from person to person. And that is a good thing. Bernstein (1996, p. 105) writes:

> If everyone valued every risk in precisely the same way, many risky opportunities would be passed up. ... Where one sees sunshine, the other sees a thunderstorm. Without the venturesome, the world would turn a lot more slowly. Think of what life would be like if everyone were phobic about lightning, flying in airplanes, or investing in start-up companies. We are indeed fortunate that human beings differ in their appetite for risk.

Reflection

Bernoulli provides this example in his famous article: two men, each worth 100 ducats (about $4000), decide to play a fair game (i.e. a game where the expectation is the same for both players) based on tossing coins, in which there is a 50–50 probability of winning or losing. Each man bets 50 ducats on the throw, which means that he has an equal probability of ending up worth 150 ducats or of ending up worth only 50 ducats. Would a rational player play such a game?

The expectation is 100 ducats for each player, whether they decide to play or not. But most people would find this play unattractive. Losing 50 ducats hurts more than gaining 50 ducats pleases the winner. There is an asymmetry in the utilities. The best decision for both is to refuse to play the game.

Risk-averse behaviour

Economists and psychologists refer to the players as risk-averse. They dislike the negative outcomes more than the weight given by the expected value. The use of the term 'risk averse' is based on a concept of risk that is linked to uncertainties more than expected values (see Chapter 4). Hence this terminology is in conflict with the idea of seeing risk as the expected value.

Let us return to the Russian roulette game described above. Imagine that you were given a choice between a gift of $0.5 million for certain or an opportunity to play the game with uncertain outcomes. The gamble has an expectation equal to $1 million. Risk-averse people will choose the gift over the gamble. As the possible loss is so large ($24 million), they would probably prefer any gift (even a fixed loss) instead of accepting the game. The minimum gift you would require is referred to as the certainty equivalent. A person is risk-averse if the certainty equivalent is less than the expected value. Different people would be risk-averse to different degrees. This degree is expressed by the certainty equivalent. How high (low) would the gift have to go before you would prefer the game to the gift?

For the above examples, most people would show a risk-averse attitude. A risk seeker would have a higher certainty equivalent than the expected value. He or she values the probability of winning to be so great that (s)he would prefer to play the game instead of receiving the gift of say $1.2 million.

A portfolio perspective

But is it not more rational to be risk-neutral, that is, letting the certainty equivalent be equal to the expected value? Say that you represent an enterprise with many activities and you are offered the Russian roulette type of game. The enterprise is huge, with a turnover of billions of dollars and hundreds of large projects. In such a case the enterprise management would probably accept the game, as the expectation is positive. The argument is that when considering many such games (projects) the expected value would be a good indication of the actual outcome of the total value of the games (projects).

To illustrate this, say that the portfolio of projects comprises $n = 100$ projects and each project is of the Russian roulette type, that is, the probability of losing $24 million is 1/6, and the probability of winning $6 million is 5/6. For each project the expected gain equals $1 million and hence the expected average gain for the 100 projects is $1 million. Looking at all the projects we would predict $1 million per project, but the actual gain could be higher or lower. There is a probability that we lose hundreds of million of dollars, but the probability is rather low. In theory, all projects could result in a loss of $24 million, adding up to loss of $2400 million. Assuming that all the n projects are independent of each other, the probability of this extreme result is $(1/6)^{100}$, which is an extremely small number; it is negligible. It is, however, quite likely that we end up with a loss, that is, negative average gain. To compute this probability we make use of the central limit theorem, expressing the fact that the probability distribution of the average value can be accurately approximated by the normal (Gaussian) probability curve. As shown in Table 1.1, the probability that the average gain is less than zero

Table 1.1 Probability distribution for the average gain when $n = 100$.

X	Probability that the average gain is less than x
−2	0.004
−1	0.04
0	0.19
1	0.50
2	0.81
3	0.96
4	0.996

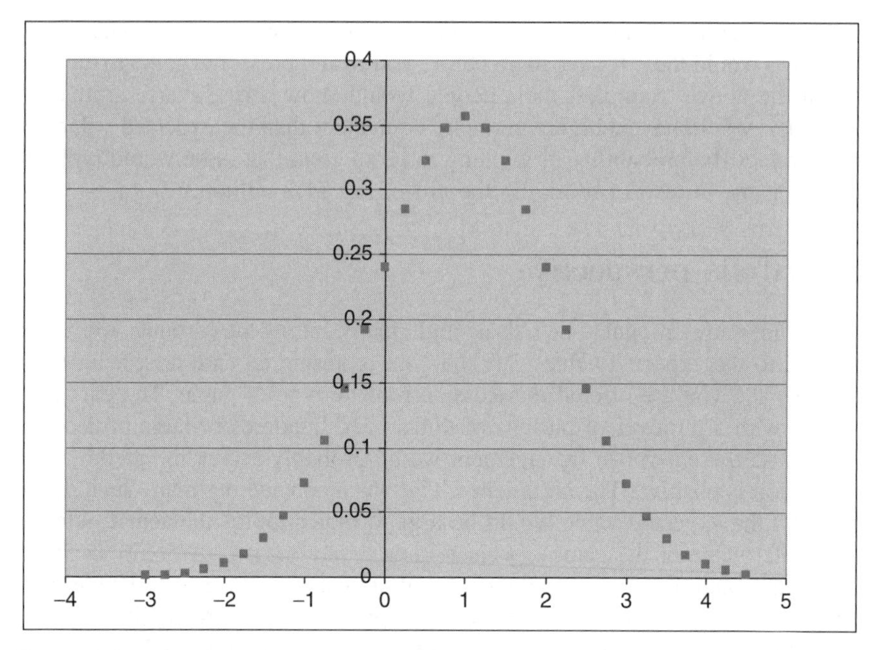

Figure 1.1 The Gaussian curve for the average gain in millions of dollars for the case $n = 100$. The area under the curve from a point a to a point b on the x-axis represents the probability that the gain takes a value in this interval.

equals approximately 0.20, assuming that all projects are independent. Figure 1.1 shows the Gaussian curve for the average gain. The probability that the average gain will take a value lower than any specific number is equal to the area below the curve. The integral of the total curve is 1. Hence, the probability is 0.50 that the average gain is less than 1, and 0.50 that the average gain exceeds 1. Table 1.1 provides a summary of some specific probabilities.

The central limit theorem has an interesting history, as Tijms (2007, p. 162) describes:

The first version of this theorem was postulated by the French-born mathematician Abraham de Moivre, who, in a remarkable article published in 1733, used the normal distribution to approximate the distribution of the number of heads resulting from many tosses of a fair coin. This finding was far ahead of its time, and was nearly forgotten until the famous French mathematician Pierre-Simon Laplace rescued it from obscurity in his monumental work *Théorie Analytique des Probabilités*, which was published in 1812. Laplace expanded De Moivre's finding by approximating the binomial distribution with the normal distribution. But as with De Moivre, Laplace's finding received little attention in his own time. It was not until the nineteenth century was at an end that the importance of the central limit theorem was discerned, when, in 1901, Russian mathematician Aleksandr Lyapunov defined it in general terms and proved precisely how it worked mathematically. Nowadays, the central limit theorem is considered to be the unofficial sovereign of probability theory.

Calculations of the figures in Table 1.1

The calculations are based on the expected value, which equals 1, and the variance, which is a measure of the spread of the distribution relative to the expected value. For one project, the variance equals

$$(-24 - 1)^2 \times 1/6 + (6 - 1)^2 \times 5/6 = 25^2 \times 1/6 + 5^2 \times 5/6 = 750/6 = 125.$$

We see that the variance is computed by squaring the difference between a specific outcome and the expected value, multiplying the result by the probability of this outcome, and then summing over the possible outcomes. If X denotes the outcome we denote by $\mathrm{E}[X]$ the expected value of X, and $\mathrm{Var}[X]$ the variance of X. Formally, we have $\mathrm{Var}[X] = E([X - EX])^2$.

The square root of the variance is called the standard deviation of X, and is denoted $\mathrm{SD}[X]$. For this example we obtain $SD[X] = 11.2$. The variance of a sum of independent quantities equals the sum of the individual variances. Let Y denote the total gain for the 100 projects. Then the variance of Y, $\mathrm{Var}[Y]$, equals 12 500.

The central limit theorem states that

$$\mathrm{P}(Y/n \leq x) \approx \Phi(\sqrt{n}(x - \mathrm{E}X)/\mathrm{SD}[X]),$$

where $\sqrt{n}$ equals the square root of n and Φ is the probability distribution of the standard normal distribution with expectation 0 and variance 1. The approximation $\approx$ produces an accurate result for large n, typically larger than 30. The application of this formula gives

$$\mathrm{P}(Y/n < 0) \approx \Phi(-10/11.2) = \Phi(-0.89) = 0.19,$$

using a statistical table for the Φ function. The standard deviation for Y/n equals $SD[X]]/\sqrt{n} = 1.12$.

We observe that the expected value is a more informative quantity when looking at 100 projects of this form than looking at one in isolation. The prediction is the same, \$1 million per project, but the uncertainties have been reduced. And if we increase the number of projects the uncertainties are further reduced.

Table 1.2 Probability distribution for the average gain when $n = 1000$.

X	Probability that the average gain is less than x
−0.25	0.0002
0	0.002
0.5	0.08
1	0.50
1.5	0.92
2	0.998

Say that we consider $n = 1000$ projects. Then we obtain results as in Table 1.2 and Figure 1.2. We see that the probability of a loss in this case is reduced to 0.2%. The outcome would with high probability be a gain close to \$1 million. The uncertainties are small. Increasing n even further would give stronger and stronger concentration of the probability mass around 1. This can be illustrated by the variance or the standard deviation. For the above example the standard deviation of the average gain, $\mathrm{SD}[Y/n]$, equals 1.12 in case $n = 100$ and 0.35 when $n = 1000$. As the number of projects increases, the variance and

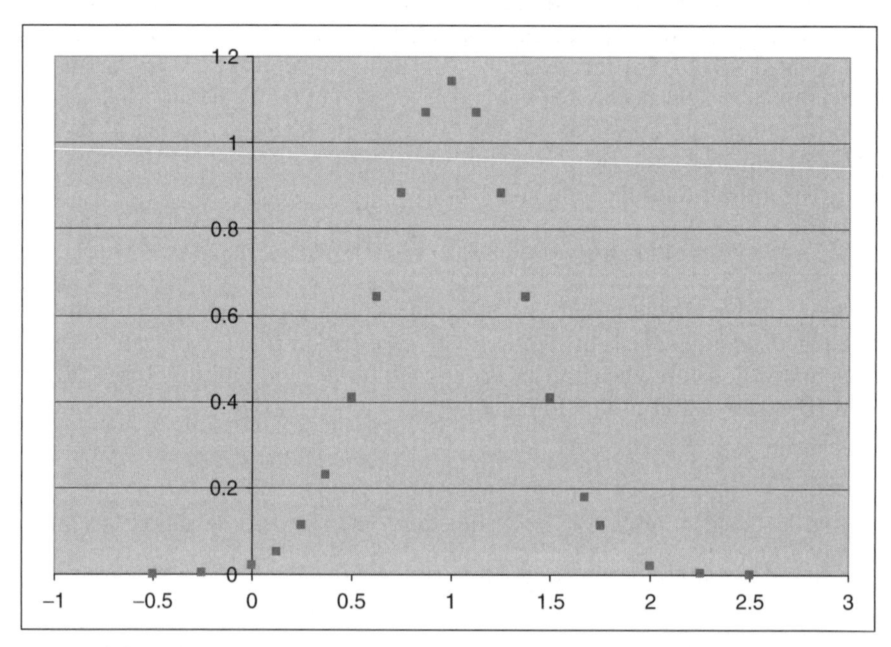

Figure 1.2 The Gaussian curve for the average gain in millions of dollars for the case n = 1000. The area under the curve from a to b represents the probability that the gain takes a value in this interval.

Table 1.3 Probability distribution for the average gain when $n = 10\,000$.

X	Probability that the average gain is less than x
0.60	0.0002
0.90	0.19
1	0.50
1.1	0.81
1.4	0.9998

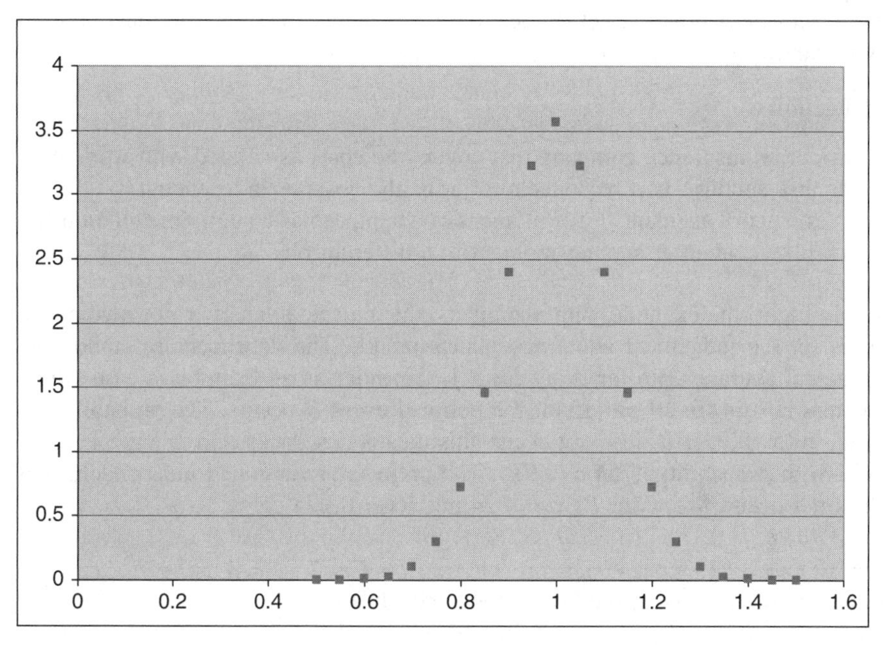

Figure 1.3 The Gaussian curve for the average gain in millions of dollars for the case n = 10 000. The area under the curve from a to b represents the probability that the gain takes a value in this interval.

the standard deviation decrease. When n becomes several thousands, the variance and the standard deviation become negligible and the average gain is close to the expected value 1. See Table 1.3 and Figure 1.3 which present the result for $n = 10\,000$.

Dependencies

The above analysis is based on the assumption that the projects are independent. The law of large numbers and the central limit theorem require that this assumption is met for the results to hold true. But what does this mean and to what extent is this a realistic assumption? Consider two games (projects), and let X_1 and X_2 denote the gains for games 1 and 2, respectively. These games are independent if the probability distribution of X_2 is not dependent on X_1, that is, the probability that the outcome of X_2 turns out to be -24 or 6 is not dependent on whether the outcome of X_1 is -24 or 6. For the Russian roulette type game independence is a reasonable assumption under appropriate experimental conditions. For real-life projects the situation is, however, more complex. If you know that project 1 has resulted in a loss, this may provide information also about project 2. There could be a common cause influencing both projects negatively or positively. Think about an increase in the oil price or a political event that influences the whole market. Hence, the independence assumption must be used with care.

Reflection

Consider an insurance company that covers the costs associated with work accidents in a country. Is it reasonable to judge the costs as independent?

Yes, a work accident at one moment in one place has a negligible relationship to a work accident at another moment in a different place. □

Returning to the example with n projects, the expected value cannot be the sole basis for the judgement about acceptance or not. The dependencies could give an actual average gain far away from 1. Consider as an example a case where the loss is -24 for all projects if the political event B occurs. The probability of B is set to 10%. If B does not occur, this means that the loss in one game, X, is -24 with probability $4/54 = 0.074$. The projects are assumed independent if B does not occur. We write $P(X = -24|\text{not } B) = 0.074$.

This is seen by using the law of total probability:

$$\begin{aligned} \mathrm{P}(X = -24) &= \mathrm{P}(X = -24|B) \cdot \mathrm{P}(B) + \mathrm{P}(X = -24|\text{not } B) \cdot \mathrm{P}(\text{not } B) \\ &= 1 \cdot 0.1 + \mathrm{P}(X = -24|\text{not } B) \cdot 0.9 = 1/6. \end{aligned}$$

The expected gain and variance given that B does not occur equal

$$\mathrm{E}[X|\text{not } B] = -24 \cdot (4/54) + 6 \cdot (50/54) = 204/54 = 3.78,$$

$$\mathrm{Var}[X|\text{not } B] = (-24 - 3.78)^2 \cdot (4/54) + (6 - 3.78)^2 \cdot (50/54) = 61.7.$$

Hence for one particular project we have the same probability distribution: the possible outcomes are 6 and -24, with probabilities 5/6 and 1/6, respectively. The projects are, however, not independent. The variance of the average gain, $\mathrm{Var}[Y/n]$, does not converge to zero as in the independent case.

To see this, we first note that

$$\text{Var}[Y] = \text{E}(Y - \text{E}Y])^2 = \text{E}[(Y - \text{E}Y])^2|B] \cdot 0.1 + \text{E}[(Y - \text{E}Y])^2|\text{not } B] \cdot 0.9$$
$$\geq (25n)^2 \cdot 0.1 = 62.5n^2.$$

Consequently $\text{Var}[Y/n] \geq 62.5$ and the desired conclusion is proved.

Hence, for large n the probability distribution of the average gain Y/n takes the following form:

- There is a probability of 0.1 that Y/n equals -24.
- There is a probability close to 0.90 that Y/n is in an interval close to 3.78.

If for example $n = 10\,000$, the interval is [3.6, 3.9]. This interval is computed by using the fact that if B does not occur, the expected value and standard deviation (SD) of Y/n equals 3.78 and $\sqrt{61.7/n}$, respectively. Using the Gaussian approximation, the interval $3.78 \pm 1.64 \cdot \text{SD}$ has a 90% probability.

We can conclude that there is a rather high probability of a large loss even if the number of projects is large. The dependence causes the average gain not to converge to the expected value 1.

Should not risk as a concept explicitly reflect this probability of a loss equal to -24? The expected value 1 is not very informative in this case as the distribution has two peaks, -24 and 3.78. Often in real life we may have many such peaks, but the probabilities could be rather small. Other definitions of risk do, however, incorporate this type of distribution, as we will see in the coming chapters.

Different distributions. Extreme observations

Above we have considered projects that are similar: they have the same distribution. In practice we always have different types of projects and some could be very large. To illustrate this, say that we have one project where the possible outcomes are -2400 and 600 and not -24 and 6. The expectation is thus 100 for this project. Then it is obvious that the outcome of this project dominates the total value of the portfolio. The law of large numbers and the central limit theorem cannot be applied. See Figure 1.4 which shows the case with $n = 100$ standard projects with outcomes -24 and 6 and one project with the extreme outcomes -2400 and 600. The probabilities are the same, 1/6 and 5/6, respectively. We observe that the distribution has two peaks, dominated by the extreme project. There is a probability of 1/6 of a negative outcome. If this occurs, the average gain is reduced to about -23. If the extreme project gives a positive result, the average gain is increased to about 7. The 100 standard projects are not sufficiently many to dominate the total portfolio. The computations of the

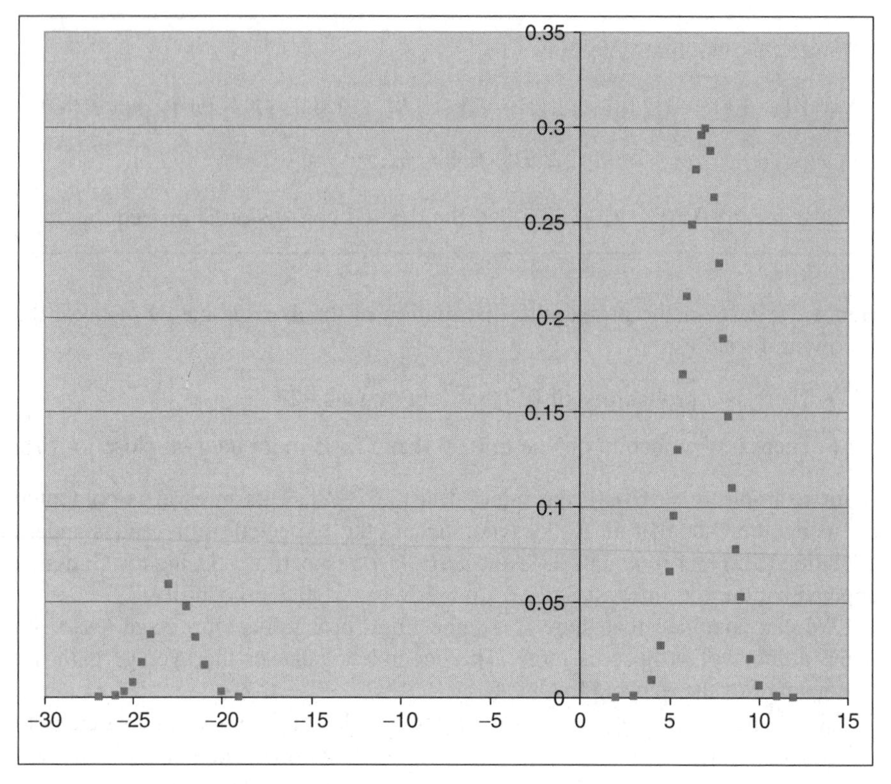

Figure 1.4 The probability distribution for the average gain for n = 100 standard projects and one project with outcomes −2400 and 600. The area under the curve from a to b represents the probability that the gain takes a value in this interval.

numbers in Figure 1.4 are based on the following arguments: If Y denotes the sum of the gains of the $n = 100$ standard projects and Y_L the gain from the extreme project, the task is to compute $P((Y + Y_L)/101 \leq y)$. But this probability can be written as

$$\begin{aligned} \mathrm{P}(Y \leq 101y - Y_L) &= \mathrm{P}(Y \leq 101y + 2400 | Y_L = -2400) \cdot 1/6 \\ &\quad + \mathrm{P}(Y \leq 101y - 600 | Y_L = 600) \cdot 5/6 \\ &= \mathrm{P}(Y/100 \leq 1.01y + 24) \cdot 1/6 \\ &\quad + \mathrm{P}(Y/100 \leq 1.01y - 6) \cdot 5/6, \end{aligned}$$

and the problem is of the standard form analysed earlier for Y. We have assumed that project gains are independent.

If the number of the standard projects increases, the value of the extreme project is reduced, but it is obvious that some large projects could have a significant influence on the total value of the portfolio. Figures 1.5 and 1.6 are

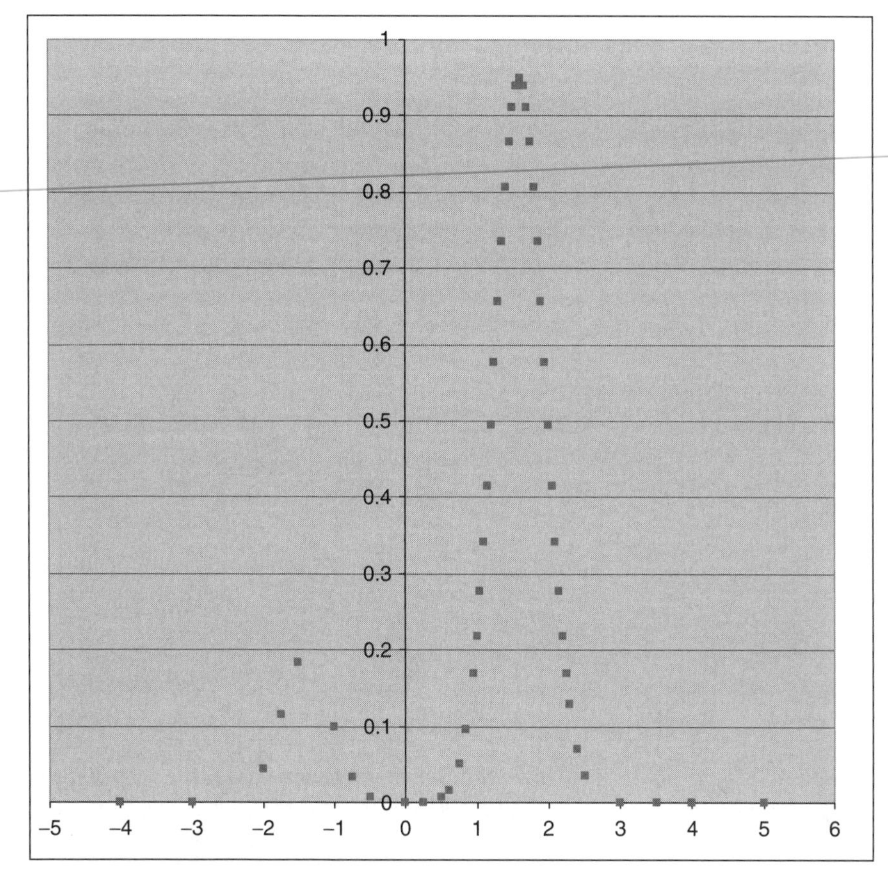

Figure 1.5 The probability distribution for the average gain for n = 1000 and one project with outcomes −2400 and 600. The area under the curve from a to b represents the probability that the gain takes a value in this interval.

similar to Figure 1.4 but $n = 1000$ and 10 000, respectively. The influence of the extreme project is reduced, but for $n = 1000$ the probability of a negative outcome is still about 1/6. However, in the case of $n = 10\,000$, the probability of a negative outcome is negligible. The probability mass is now concentrated around the expected value 1. We see that a very large number of standard projects are required to eliminate the effect of the extreme project.

Difficulties in establishing the probability distribution

In the above analysis there is no discussion about the probability distribution for each project. There is a probability of 1/6 of the negative outcome and a probability of 5/6 of the positive outcome. In practice we seldom have such a distribution available. If you run a business project or you invest your money on

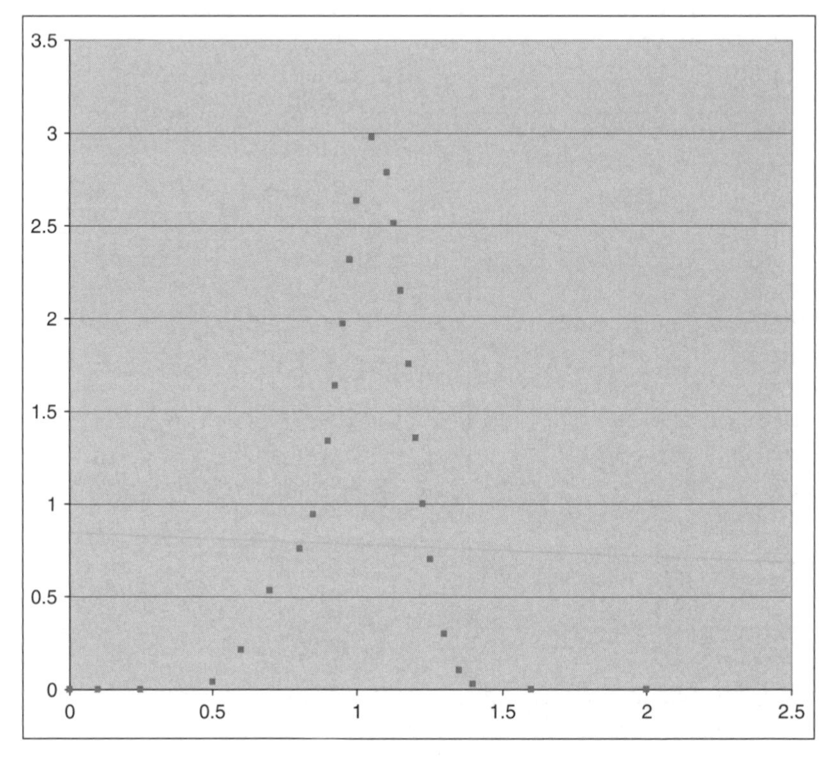

Figure 1.6 The probability distribution for the average gain for n = 10 000 and one project with outcomes −2400 and 600. The area under the curve from a to b represents the probability that the gain takes a value in this interval.

the stock market, it is not obvious how to determine the probability distributions. You may have available some historical data, but to what extent are they relevant for the future? Let us simplify our analysis by considering $n = 10\,000$ future projects and assume that the possible outcomes are −24 and 6 as above, but a probability distribution has not been specified. Historical data are on hand, for similar projects, and these show a distribution between the two possible outcomes which are 1/6 and 5/6 respectively. However, we do not know if this distribution will actually be the distribution for the future projects. We may assume that this will be the case, but we do not know for sure. There are uncertainties – surprises may occur. Under the assumption that the future will be as history indicates, the results are shown by Table 1.3 and Figure 1.3. One may argue that such an assumption is reasonable and that the data are the best information available, but we have to acknowledge that not all uncertainties have been revealed by the numbers shown by Table 1.3 and Figure 1.3. Should not this uncertainty factor be considered an element of risk? We will discuss this in detail in the coming chapter.

We end this chapter with two general reflections on the use of expected values in risk management.

Reflection

In his paper from 1738 referred to above, Daniel Bernoulli presented and discussed the so-called St Petersburg paradox. This problem was first suggested by his cousin Nicolaus Bernoulli. It is based on a casino game where a fair coin is tossed successively until the moment that heads appears for the first time. The casino payoff is 2 ducats if heads comes up on the first toss, 4 ducats if heads turns up for the first time in the second toss, and so on. In general the payoff is 2^n ducats. Thus, with each additional toss, the payoff is doubled. How much should the casino require the player to stake such that, over the long run, the game will not be a losing endeavour for the casino? (Tijms, 2007; Bernstein, 1996).

The answer is that the casino owners should not allow this game to be played, whatever amount a player is willing to stake, as the expected value of the casino payoff for one player is an infinitely large number of ducats. To see that the expected value exceeds every conceivable large value, note first that the probability of getting heads in the first toss is $1/2$, the probability of getting heads for the first time at the second toss equals $1/2 \times 1/2$, and so on and the probability of getting heads for the first time at the nth toss equals $(1/2)^n$. The expected value of the game for one player is thus

$$1/2 \times 2 + (1/2)^2 \times 4 + \ldots + (1/2)^n \times 2^n + \ldots.$$

In this infinite series, a figure of 1 ducat is added to the sum each time, and consequently the sum exceeds any fixed number.

In Bernoulli's day a heated discussion grew up around this problem. Some indicated that the probabilistic analysis was wrong. But no, the mathematics is correct. However, the model is not a good description of the real world. The model tacitly assumes that the casino is always in a position to pay out, whatever happens, even in the case that heads shows up for the first time after say 30 tosses where the payoff is larger than \$1000 million. So a more realistic model is developed if the casino can only pay out up to a limited amount. Then the expected payoff is limited and can be calculated. See Tijms (2007). □

Reflection

Willis (2007) defines terrorism risk as expected damage, as mentioned above. In view of the discussion in this chapter, why is such a perspective problematic?

For terrorism risk, the possible consequences could be extreme with millions of possible fatalities, and consequently the expectation, even on a national and international level, would produce poor predictions of the actual damages and losses. In addition, the uncertainties in underlying phenomena and processes are large. It is extremely difficult to predict when the next attack will occur and what form it will take. The historical data obviously provide limited information. Any assigned probability distribution is likely to deviate strongly from the observed loss distribution.

Summary

These are the two main issues that we should reflect on:

- Is the concept of risk captured by the expected value?
- Should decisions involving risk be based on expected values?

For both questions, the answer is in general *no*. The problem is that the expected value could deviate strongly from the actual outcomes. There are two main reasons for this:

1. The consequences or outcomes could be so extreme that the average of a large population of activities is dominated by these extreme outcomes.
2. The probability distribution could deviate strongly from the future observed outcome distribution.

In gamble-like situations of repeated experiments, the expected value would provide good predictions of the actual future quantities studied, but not often in other situations.

References

Bernstein, P.L. (1996) *Against the Gods: The Remarkable Story of Risk*, John Wiley & Sons, Inc., New York.

Floodcite (2006) Guidelines for Socio-economic Flood Damage Evaluation, Report T9-06-01. February. Floodcite.

HM Treasury (2005) Managing Risk to the Public, Appraisal guidance.

Lirer, L., Petrosino, P. and Alberico, I. (2001) Hazard assessment at volcanic fields: the Campi Flegrei case history. *Journal of Volcanology and Geothermal Research*, **112**, 53–73.

Mandel, D. (2007) Toward a Concept of Risk for Effective Military Decision Making, Defence R&D Canada, Toronto, Technical Report 2007-124. DRDC Toronto.

Tijms, H. (2007) *Understanding Probability: Chance Rules in Everyday Life*, 2nd edn, Cambridge University Press, Cambridge.

Verma, M. and Verter, V. (2007) Railroad transportation of dangerous goods: population exposure to airborne toxins. *Computers and Operations Research*, **34**, 1287–1303.

Willis, H.H. (2007) Guiding resource allocations based on terrorism risk. *Risk Analysis*, **27** (3), 597–606.

Further reading

Bernstein, P.L. (1996) Against the gods: *The remarkable story of risk*, Wiley, New York.

Tijms, H. (2007) *Understanding Probability: Chance Rules in Everyday Life*, 2nd ed. Cambridge University Press, Cambridge.

2

Risk is a probability or probability distribution

Let us return to the Russian roulette game introduced in Chapter 1, but let us replace the revolver with a die. Say that a person (John) on the street offers you one game with the following payoffs:

If the die shows a 6, you lose $24 000, and otherwise you win $6000.

John selects the die but you throw the die. Would you play the game? Many people would do if they were sure that the die were normal. In that case the probability of wining $6000 would be 5/6 and the probability of losing $24 000 equal to 1/6. The expected value is equal to $1000. However, the probability distribution is based on the assumption of a normal (fair) die, but you cannot be sure that the die used is fair. So your probability distribution is not providing you with all the information you need for making your decision. The risk is certainly more than the probabilities produced.

To reflect this risk factor you assign a probability equal to 0.90 that John is not using a fair die. You believe John is cheating as the game seems very much in your favour.

Then the probability that you win $6000 equals $5/6 \times 0.10$, assuming that if John is not using a fair die there is zero probability that you win. Hence the probability of winning equals $1/12 = 0.083$, that is, about 8%, and the game is obviously not particularly attractive to you. The expected value equals $-21\,500$.

Common risk definitions based on probability

Does this mean then that the concept of risk is fully captured by the probabilities? Yes, conclude many risk researchers and analysts. They define risk solely through probabilities. Here are some examples:

Misconceptions of Risk T. Aven

(a) Risk is the probability of an undesirable event (see Campbell, 2005).

(b) Risk is the probability of an adverse outcome (Graham and Weiner, 1995).

(c) Risk is a measure of the probability and severity of adverse effects (Lowrance, 1976).

(d) Risk is the combination of probability of an event and its consequences (ISO, 2002).

(e) Risk is probability and consequence, more specifically risk is equal to the triplet (s_i, p_i, c_i), where s_i is the ith scenario, p_i is the probability of that scenario and c_i is the consequence of the ith scenario, $i = 1, 2, \ldots, N$ (Kaplan and Garrick, 1981; Kaplan, 1991).

Let us look more closely into these definitions. The first associates risk with an undesirable event which, in the example, would be losing money, that is, \$24 000. Risk equals the probability that you lose this amount of money. However, this is obviously an inadequate description of risk as you do not relate this probability to the other possible outcomes. In this case there are only two possible outcomes, but in general there could be many and the restriction to one undesirable event means that the extent or significance of the loss is not reflected. If we consider the undesirable event *machine failure*, the consequences or outcomes could range from negligible to catastrophic depending on the availability and performance of a set of barriers, as well as the extent of exposure of human lives, and other objects that humans value. In our die example we would relate the loss of \$24 000 to winning \$6000, which is also an argument for not restricting risk to negative outcomes only; refer to the discussion in Chapter 9.

Although definition (a) is not suitable as a general definition of risk, it is not difficult to find situations where it is an informative risk measure (index). Consider, for example nuclear power plants and the risk related to a core melt. Then the probability of a core melt captures an essential element of risk as any severe consequence for human lives and the environment would be a result of a core melt. We know that the consequences of a core melt could be dramatic, so the focus is on the possible occurrence of this event (Borgonovo and Apostolakis, 2001). Another example is the risk of getting a specific disease, for example cancer. In health care it is common to talk about cancer risk, understood as the probability of getting cancer.

Obviously the probabilities of the events core melt and getting cancer provide much risk-related information. However, the risk concept should also capture the consequences of the core melt or the cancer. In one case, the consequences of getting cancer could be rather small, in other cases life-threatening. Risk should also cover this aspect, and the intention of definitions (c)–(e) is to do this. Definition (b) looks similar to (a) but may also be interpreted as the probability distribution of an adverse outcome, that is, the probability associated with all relevant adverse outcomes. Consider the cancer example and assume that the consequences can be classified into four categories:

- C_1: The patient recovers from the disease within 1 year
- C_2: The patient recovers from the disease later
- C_3: The patient dies due to the cancer during the course of 5 years
- C_4: The patient dies due to the cancer later.

Suppose that the patient has cancer (denote this event by A) and the following probabilities have been specified:

$$\mathrm{P}(C_1|A) = 0.50, \quad \mathrm{P}(C_2|A) = 0.10, \quad \mathrm{P}(C_3|A) = 0.30, \quad \mathrm{P}(C_4|A) = 0.10,$$

that is, a probability of 50% that the patient recovers from the disease within 1 year, and so on. These probabilities are conditioned on the occurrence of the undesirable event A (cancer). If the probability of cancer during a specific period of time is 0.01, say, the unconditional probabilities become

$$\mathrm{P}(A\&C_1) = 0.05, \quad \mathrm{P}(A\&C_2) = 0.01, \quad \mathrm{P}(A\&C_3) = 0.03, \quad \mathrm{P}(A\&C_4) = 0.01.$$

Hence, the probability that the person gets cancer during the specified period of time and recovers within a year is 0.05, and so on.

Definition (c) expresses essentially the same as the above probabilities. Severity is just a way of characterizing the consequences and refers to intensity, size, extension, scope and other potential measures of magnitude, and relates to something that humans value (lives, the environment, money, etc.). Gains and losses, for example expressed in terms of money or the number of fatalities, are ways of defining the severity of the consequences.

Definition (d) can also be interpreted along the same lines, although the formulation seems to restrict the probability dimension to the probability of the event, the cancer in our example. However, probabilities also need to be linked to the consequences as was shown in the above example.

Definition (e) has a focus on scenarios, which in our example are:

- s_1: The patient gets cancer and recovers from the disease within 1 year
- s_2: The patient gets cancer and recovers from the disease later
- s_3: The patient gets cancer and dies due to the cancer during the course of 5 years
- s_4: The patient gets cancer and dies due to the cancer later.

Associated with each scenario there is a probability $p_i = \mathrm{P}(s_i)$. In this example the consequences are described through the scenarios. In other cases we may, for example classify the consequences into severity categories reflecting the number of fatalities.

How to specify or estimate the probability distribution

The above analysis requires that probabilities and probability distributions have been specified or estimated. How is this done? Basically we may distinguish between four approaches:

1. Direct use of historical data
2. Direct assignments or estimates
3. Use of standard probability distributions, such as the Poisson distribution and the normal distribution
4. Use of detailed modelling of phenomena and processes, for example using event trees, fault trees and Bayesian belief networks. The analysis may use Monte Carlo simulation to obtain the probabilities based on the models.

For each of these approaches a number of principles and methods exist. It is, however, beyond the scope of this book to present and discuss all these principles and methods. Nonetheless, many of the coming chapters address issues related to these principles and methods. For example Chapter 10 discusses the use of historical data to determine risk and Chapter 13 studies the use of models in risk analysis. Key references to the principles and methods are Haimes (2004), Vose (2008) and Aven (2003, 2008).

Using the above definitions, risk is restricted to the probabilistic world. However, such a perspective can be challenged. A probability does not capture all aspects of concern. Let us return to the die example. Obviously the first assessment without considerations of John as a cheater lacked one important aspect. But is the extended assessment, which includes an assignment of the probability that the die is not fair, also too narrow? Does the assigned probability of 0.90 that the die is not fair capture the missing aspect? The figure 0.90 seems rather arbitrary and is not derived very rigorously. Would it not be reasonable to require that the probabilities are objective – in the sense that they are not dependent on the assessor?

To be able to respond to these questions we need to clarify what we mean by a probability.

The meaning of a probability

The probability of throwing a 2 with a fair die is 1/6. This is intuitively obvious as there are six outcomes and by symmetry they should have the same probability. But now consider throwing a drawing pin, as a result of which the pin is either up or down. What is the probability that the pin is up? In this case we cannot refer to outcomes that are equally probable, but we can perform a number of trials and see how often the pin shows up and how often it shows down. Say that we perform 100 trials and the pin shows up in 70 cases. Then the probability sought must be close to 0.70. We associate the probability with the success rate, that is, the relative frequency of up occurrences. This is often referred to as the relative frequency interpretation of probability.

Relative frequency interpretation

To be precise, a frequentist probability (i.e. a probability interpreted as a relative frequency) is defined by the fraction of 'successes' if the experiment is repeated over and over again an infinite number of times. The probability is unknown, and needs to be estimated. Hence 0.70 is not the probability but an estimate of the true underlying probability. To make this clear, let p denote the probability that the pin is up. Then 0.70 is an estimate of p. We denote the estimate by p^* and consequently we have $p^* = 0.70$.

If this interpretation is adopted we have to take into account that the estimates could be more or less close relative to the underlying true probability. We say there is estimation uncertainty. Textbooks in statistics study these uncertainties using measures such as variance and confidence intervals. Let us look at an example.

Example. Estimation Error

Let X denote the number of times the pin shows up when throwing the pin $n = 1000$ times. Then $p^* = X/1000$ is an estimator of p. In the above example we have made the observations and computed an estimate $p^* = 0.70$. The estimator p^* (before we observe the outcome of the experiment) has an expected value equal to p, that is,

$$\mathrm{E}[p^*] = \mathrm{E}[X/1000] = p.$$

The expected number of successes in one throw, $\mathrm{E}[X_1]$, equals p as the result of one throw could be either 1 or 0, with probabilities p and $1 - p$ respectively. Hence $\mathrm{E}[X/1000] = 1000 \cdot p/1000 = p$.

The law of large numbers says that p^* converges to p as the number of throws n goes to infinity. However, for $n = 1000$ as in our example the estimate may deviate from the true value of p. The variance is used as a measure to express how large deviations we can expect. The variance of the estimator is a measure of spread of the estimator compared to the expected value p. If the variance is small, the estimator will be close to p with a high probability. The variance of the estimator p^* equals

$$\mathrm{Var}[p^*] = \mathrm{Var}[X/n] = \mathrm{Var}[X]/n^2 = n\mathrm{Var}[X_1]/n^2 = p(1 - p)/n$$

as $\text{Var}[X_1] = \text{E}(X_1 - p)^2 = (1-p)^2 p + (0-p)^2(1-p) = p(1-p)$.

We see that if n increases, the variance of the estimator converges to zero. Hence we can conclude that the estimation error is small. More explicitly, we have the following inequality: $\text{P}(|p^* - p| \geq d) \leq \text{Var}[p^*]/d^2$. This inequality, which is referred to as the Chebyshev's inequality, states that the probability that the estimation error $|p^* - p|$ is greater than a small number d (say, 0.05), is less than $\text{Var}[p^*]/d^2$. But $\text{Var}[p^*]/d^2 \leq 1/(4nd^2)$ as $p(1-p) \leq$ $1/4$. Hence

$$\text{P}(|p^* - p| \geq 0.05) \leq 0.10,$$

that is, the probability that the estimation error exceeds 0.05 is less than 10%. In other words, we are confident that the estimator we use in this case with $n = 1000$ throws produces estimates within a 0.05 error relative to the true underlying probability. We cannot say that p lies in the interval [0.65, 0.75] with minimum 90% probability, as it is meaningless to speak of probabilities of fixed numbers such as 0.65 and 0.75. The probability statements must be linked to the unknown quantities (referred to as stochastic variables) before the results of the experiment are put into the formula.

A confidence interval with confidence 90% (typical confidence intervals are 90%, 95% and 99%) is a similar type of interval. It is an interval $[p_{\text{L}}, p_{\text{H}}]$, where p_{L} and p_{H} are stochastic variables depending on the observations X such that

$$\text{P}(p_{\text{L}} \leq p \leq p_{\text{H}}) = 0.90.$$

Hence, the probability that $[p_{\text{L}}, p_{\text{H}}]$ includes the true p equals 90%. In this case such an interval is given by

$$[p^* - 1.64 \times \text{SD}[p^*], \quad p^* + 1.64 \times \text{SD}[p^*]],$$

where $\text{SD}[p^*]$ is the standard deviation of p^* which equals the square root of $p(1-p)/n$. As p is unknown we need to replace p by p^* and this gives the interval [0.68, 0.72] for this particular case with $n = 1000$ and $p^* = 0.70$. This interval is an approximate 90% confidence interval as it is based on the use of two approximations; the central limit theorem (see Chapter 1) and the use of p^* instead of p in the expression for the standard deviation of p^*. The approximations are, however, good as n is large.

The great mathematician and probabilist Jacob Bernoulli (1654–1705) faced similar precision problems when he tried to determine probabilities of specific events. He required an accuracy of 1/1000 and used the term moral certainty about this accuracy. He found that a total number of 25 500 trials were required to provide this accuracy. This number seemed too large for Jacob and he had trouble digesting it. He worked on a book project on the topic, *Ars Conjectandi*, but did not finalize it. Many researchers have concluded that Jacob was not convinced by his examples. He was not able to determine probabilities with the necessary precision. On 3 October 1703, he wrote to Leibniz: 'I have completed the larger part of my book, but the most important part is missing, where I show how the basics of *Ars Conjectandi* can be applied to civil, moral and economic matters' (Polasek, 2000).

The problem with this way of thinking, the search for the accurate specification of an underlying correct probability, is the need for experimental repetitions. The analysis presumes that we can establish a large population of similar activities (in the example throws of the pin), that is, the experiment can be repeated over and over again under similar conditions. In practice this is not easily achieved when we leave the world of gambling. Think of the die example where you suspect John to be cheating and you assign a probability of 0.90 that John is not using a fair die. Is this an estimate of a frequentist probability? In that case there exists an underlying probability that John is cheating, call it p. To define p we must think of an infinitely large population of similar situations where you are offered similar games by similar persons like John, and p equals the fraction of situations where the persons are cheating. Defining the 'experimental conditions' for such situations is, however, difficult. This is a unique situation where you are facing this particular offer by John. There is no meaning in specifying similar situations, which is required to define p. Consequently, such a p does not exist and talking about estimation of p has no meaning.

Reflection

Think of the next World Cup in football (soccer). A person refers to the probability that England qualifies and wins the tournament. What does this mean if he/she is adopting the relative frequency perspective?

According to the relative frequency perspective, we interpret the probability as the fraction of times England will win when repeating the qualification and tournament infinitely. But it would not be possible to perform such experiments. Think of one particular match, England v. Norway. Try to think of a new match with the same players, independent of the first match. It does not work. We simply cannot define the population that is required to define a frequentist probability in this case. □

For such situations we may, however, use so-called subjective probabilities, or, as we prefer to call them, knowledge-based probabilities.

Knowledge-based probabilities (subjective probabilities)

Let us again return to the die example where you suspect John to be cheating and you assign a probability equal to 0.90 that John is not using a fair die. This is a knowledge-based probability. It means that you compare the uncertainty (and likelihood) about John cheating with drawing at random a red ball from an urn having 10 balls of which 9 are red. You have the same uncertainty (likelihood) that John is a cheater as drawing a red ball. Probability is a measure of uncertainty about events and outcomes (consequences), seen through the eyes of the assessor and based on the available knowledge. It is a probability in the light of current knowledge (Lindley, 2006, p. 43). Objective probabilities do not exist. If K denotes the knowledge the probability is based on (we refer to it as the background knowledge), we can write

$$\mathrm{P}(\text{John cheats}|K) = 0.90.$$

This represents a completely different way of thinking compared to the relative frequency perspective. Often we omit the K in the probability statements as the background knowledge is tacitly understood and remains unchanged throughout the analysis.

A knowledge-based probability can always be specified. There is no need to define an underlying population of similar situations. Risk analysts assign a probability of a fatal accident occurring next year, the probability of a terrorist attack occurring, and so on. A number can be specified and the interpretation is with reference to the urn standard.

In the literature such probabilities are called *subjective probabilities*. The subjective theory of probability was proposed independently and at about the same time by Bruno de Finetti in Italy in *Fondamenti Logici del Ragionamento Probabilistico* (1930) and Frank Ramsey in Cambridge in *The Foundations of Mathematics* (1931); see Gillies (2000). L. J. Savage expanded the idea in *The Foundations of Statistics* (Savage, 1962). A subjective probability can be given different interpretations. Among economists and decision analysts, and the earlier probability theorists, a subjective probability is linked to betting: a degree of belief reflected in the odds and stakes that the subject is willing to bet on the proposition at hand. According to this perspective the probability of the event A, $\mathrm{P}(A)$, equals the amount of money that the assigner would be willing to bet if he/she were to receive a single unit of payment if the event A occurred, and nothing otherwise (Singpurwalla, 2006).

This betting interpretation extends beyond the realm of uncertainty assessments, as it reflect the assessor's attitude to money and the gambling situation (Flage *et al.*, 2009). Consider the following example by Lindley (2006). In order to specify the probability of a nuclear accident, two gambles (events) are introduced:

- Receive 100 units of payment if the nuclear accident occurs
- Receive 100 units of payment if a favourable ball is drawn from an urn containing $100p\%$ favourable balls.

Then the value of p such that the two gambles are considered equivalent should be the probability of a nuclear accident. But receiving the payment would be trivial if the accident were to occur (the assessor might not be alive to receive it). Drawing a favourable ball, on the other hand, would affect neither the assessor nor the payment.

Compare the above way of assessing uncertainty with introducing the following two events:

- The nuclear accident occurs
- Draw a favourable ball from an urn containing $100p\%$ favourable balls.

The value of p such that the uncertainty about the two events is considered equivalent should be the probability of a nuclear accident. Now gambling for reward is not the basis for this process and thus the assessment of uncertainty is not confounded with the desirability of rewards.

We refer to this interpretation of a subjective probability as a knowledge-based probability with reference to a standard, to distinguish it from the betting interpretation. Probability is a way of measuring uncertainty and the reference is the urn standard. We may also speak about the probability as a degree of belief, using the same urn reference model.

The subjectivist interpretation of probability is often called the Bayesian standpoint, after Thomas Bayes (1702–1761) who proved a special case of what is now called Bayes' theorem. One might speculate whether Thomas Bayes would have endorsed the subjectivist interpretation of probability that is now associated with his name.

The Bayesian perspective on probability has been thoroughly discussed in the literature; see, for example Lindley (2000, 2006). It has been subject to strong criticism from many researchers for being non-scientific. There are no objective answers. Science cannot claim to possess the truth. The Bayesian response is that the probabilities reflect the knowledge available about phenomena and processes. They express the remaining uncertainties. Scientific methods are used to understand these phenomena and processes, for example related to gas explosions, but there are uncertainties. In a risk analysis we cannot predict the explosion pressure with high precision. Probabilities are used to express the uncertainties using the data and insights we have available. Uncertainties will always be present and we need a tool to describe them, and knowledge-based probabilities are an adequate tool for this purpose. Other sciences try to describe the world using laws and theories without uncertainties. But we all know that certainty can only be achieved under strong experimental conditions, not in the real world.

We need to look beyond probabilities to express risk

Again we consider the die example where you suspect John to be cheating. You assign a probability of 0.90 that John is not using a fair die. This assignment is based on a Bayesian perspective on probability. If we were to apply a relative frequency interpretation of probability, 0.90 needs to be seen as an estimate of an underlying true probability representing the fraction of situations where the person is cheating when repeating the situations over and over again. As discussed above, it is difficult to understand the meaning of such a probability, but it must be defined if this perspective is to be applied.

Instead, let us first consider a case where it is somewhat easier to define the underlying frequentist probability. A person has contracted a certain disease and the physician would like to estimate the probability that the patient will be cured by using a treatment T. The probability, denoted by p, is defined as the fraction of patients who will be cured by this treatment when considering a very large (in theory, an infinite) population of similar patients. The statistical data that the physician has available provide an estimate $p^* = 0.80$. Confidence intervals are used to express the uncertainties associated with the estimation of p. However, the estimation does not reflect the uncertainties caused by using data that are not relevant to the person analysed. The analysis simply presumes that the data are relevant, but the physician knows that this patient has characteristics that are somewhat different from the typical patient in the reference population. There is, however, not much research available that provides insight into how these characteristics would influence the treatment of the patient. Hence, there are uncertainties hidden in the assumptions made for the analysis. Clearly, risk extends beyond the probabilities produced by the analysis.

This would also be the case if knowledge-based probabilities were used. In this case the physician assigns a probability distribution for p, expressing his/her uncertainties (degrees of belief) as to where the true p lies. He/she may, for example, specify the distribution presented in Table 2.1. We see that the physician assigns, for example, a probability of 40% that p lies in the interval 0.7–0.8. For this analysis to make sense the probability p must be considered relevant to the patient, that is, the defined population generating p must comprise patients similar to the one studied.

Table 2.1 Knowledge-based probability distribution for p.

p	Probability
0.5–0.6	0.05
0.6–0.7	0.20
0.7–0.8	0.40
0.8–0.9	0.20
0.9–1.0	0.15

Using this approach, the physician can reflect that the data are more or less relevant. He/she can reflect his/her knowledge based on many years of practice. The probabilities are, however, based on some background knowledge. And this background knowledge could be poor. The physician may have limited insight into the disease, and this could cause the physician to give poor predictions of the actual number of persons that would be cured by the treatment if he/she were to treat, say, 100 persons similar to this particular patient. There is a need to look beyond his/her probability assignments. Surprises relative to the assigned probabilities may occur, and by just addressing probabilities such surprises may be overlooked. The probabilities could camouflage uncertainties.

Let us look at another example. Consider the risk, seen through the eyes of a risk analyst in the 1970s, related to future health problems for divers working on offshore petroleum projects. An assignment is to be made for the probability of a diver experiencing health problems (properly defined) during the coming 30 years as a result of diving activities. Let us assume that an assignment of 1% is made. This number is based on the knowledge available at that time. There are not strong indications that the divers will experience health problems. However, we know today that these probabilities led to poor predictions. Many divers have experienced severe health problems (Aven and Vinnem, 2007, p. 7). By restricting risk to the probability assignments alone, we see that aspects of uncertainty and risk are hidden. There is a lack of understanding about the underlying phenomena, but the probability assignments alone are not able to fully describe this status.

As a third example, consider an offshore petroleum installation where the operations management is concerned about the deterioration of some critical equipment. The maintenance discipline ensures that the deterioration will not cause safety problems. It refers to a special maintenance programme that will be implemented, which will cope with the deterioration problem. So what is the risk associated with hydrocarbon leakages caused by operational problems? Given the background information of the maintenance discipline, a 10% leakage probability (for a defined leakage size) is assigned. This number is based on relevant historical data, and in no way reflects the concern of the operation's management. The assignment assumes that the maintenance programme will be effective. But surprises could occur. Production of oil over time leads to changes in operating conditions, such as increased production of water, H_2S and CO_2 content, scaling, bacteria growth, emulsions, and so on: problems that to a large extent need to be solved by the addition of chemicals. These are all factors causing increased likelihood of corrosion, material brittleness and other conditions that may cause leakages. In making an assignment of 10% we hide an important element of uncertainty. In a risk analysis a number of such probability assignments are performed, and the hidden uncertainties could create surprising outcomes somewhere. You do not know where they will come, but they definitely could happen.

Now let us return to the die example where you suspect John to be cheating. You assigned a probability of 90% that John is cheating, and based on this you obtained a probability distribution as shown in Table 2.2. Does this distribution adequately reflect risk? You may argue that this distribution expresses the possible outcomes and your honest probabilities, so yes, it is an adequate distribution of risk. But say that you were not able or willing to specify the probability that John is cheating. You find that the number assigned is too arbitrary. It is impossible to distinguish between a probability equal to 0.7 or 0.8, or even 0.5. The numbers

Table 2.2 Probability distribution for reward in the die example.

Reward ($)	Probability
6000	0.917(= 11/12)
−24 000	0.083(= 1/12)

do not seem to have a sufficiently rigorous basis. Experts on knowledge-based probabilities refer to this phenomenon as imprecision in probability assignments (Mosleh and Bier, 1996). Would risk not exist without the specification of this probability?

Yes, the uncertainties exist whether you perform the assignment or not. Uncertainty is a more fundamental concept than probability and should be the pillar of risk. Probability is just a tool used to express the uncertainties.

Reflection

A company considers an investment of $100 million in a project. The company assesses the profit and specifies a profit probability distribution as shown in Table 2.3. Hence, there is a probability of 10% that the profit exceeds $200 million, and so on. The company management needs to make a decision about the investment in this project, and risk and uncertainties are considered important input to the decision-making. To what extent do you find that the distribution of Table 2.3 provides this information?

Table 2.3 Probability distribution for the profit.

Profit ($ million)	Probability
Over 200	0.10
100–200	0.30
0–100	0.20
−100 to 0	0.20
−200 to − 100	0.15
Under − 200	0.05

As argued above, we need to address uncertainties beyond the assigned probabilities. The probability distribution presented in Table 2.3 provides the result of an assessment carried out by a specific analysis group. It is based on a number of assumptions and suppositions, for example that the historical data available provide a good characterization of the future. The knowledge of certain phenomena could be weak, and strong assumptions have to be made. However, these could turn out to be wrong. Surprises do occur. The decision-makers have to take this into account. The assumptions and suppositions made need to be evaluated together with the probabilities. □

Reflection

To describe the risk level associated with fatal accidents a company uses an F–N curve. This curve expresses the expected number (frequency) of accidents with at least N fatalities, which is approximately equal to the probability of an accident occurring with at least N fatalities, when the probabilities are small. An example is shown in Figure 2.1. We see from the figure that there is a probability of 0.001 that a fatal accident occurs. To what extent do you find this curve sufficiently informative as a risk characterization for the company?

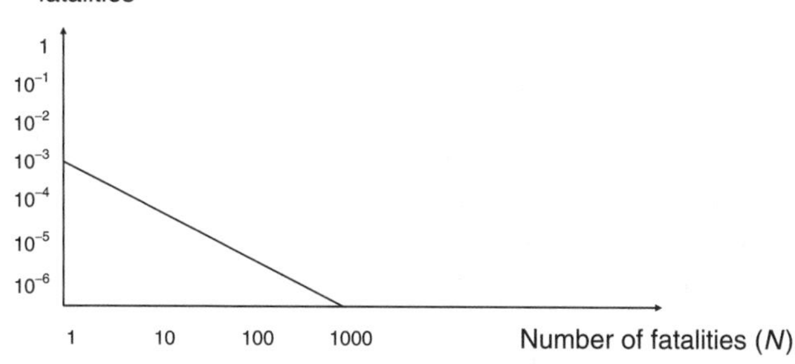

Figure 2.1 An example of an F–N curve.

We refer to the response to the previous reflection. The F–N curve represents one risk index expressed by some analysts, based on a background knowledge, which includes a set of assumptions and suppositions. □

Reflection

The uncertainties in the background knowledge are addressed by Mosleh and Bier (1996). They refer to a knowledge-based (subjective) probability $\mathrm{P}(A|X)$ which expresses the probability of the event A given a set of conditions X. As X is uncertain (it is a random variable), a probability distribution for the quantity $h(X) = \mathrm{P}(A|\mathrm{X})$ can be constructed. Thus, there is uncertainty about the random probability $\mathrm{P}(A|X)$. Would this mean that a knowledge-based probability is uncertain?

The probability is not an unknown quantity (random variable) for the analyst. To make this clear, let us summarize the setting of knowledge-based probabilities. A knowledge-based probability $\mathrm{P}(A|K)$ is conditional on the background knowledge K, and some aspects of this K can be related to X as described by Mosleh and Bier (1996). The analyst has determined to assign his/her probability based on K. If he/she finds that the uncertainty about X should be reflected, he/she would adjust the assigned probability using the law of total probability. This does not mean, however, that $P(A|K)$ is uncertain, as such a statement would presume that there exists a true probability value. The assessor needs to clarify what is uncertain and subject to the uncertainty assessment and what constitutes the background knowledge. From a theoretical point of view one may think that it is possible (and desirable) to remove all such Xs from K, but in a practical risk assessment context that is impossible. We will always base our probabilities on some type of background knowledge, and often this knowledge would not be possible to specify using quantities like X. Think of the examples above. It is not obvious what X represents. □

Reflection

If you assign a knowledge-based (subjective) probability of 0.90, is it meaningful to assign probabilities of this probability?

This issue has been much debated in the literature; see, for example Gärdenfors and Sahlin (1988) and Sahlin (1993). According to Sahlin (1993), no one would seriously dispute that we have beliefs about our beliefs. But this assertion can be questioned. As a professional risk assessor, one is trained in the process of transforming uncertainty into probabilities. If the assessor assigns a probability of an event B equal to 0.9, there is no reason why he/she should dispute his own assignment. If he/she assigns a probability equal to 0.9 based on his/her background knowledge, it is compared to drawing a red ball out of an urn consisting of 10 balls where 9 are red. He/she may experience a precision problem as discussed above, in particular when assessing events on the lower part of the probability scale. It could, for example be difficult to distinguish between numbers such as 10^{-5} and 10^{-6}. However, using knowledge-based probabilities to express uncertainties about quantities for which true values do not exist is a misuse of such probabilities. There is no rationale for using subjective probabilities in such a way. The numbers generated cannot then be given a meaningful interpretation.

Probabilities of probabilities are often referred to as second-order probabilities. In most cases the basis is frequentist probabilities, and knowledge-based probabilities are used to express the uncertainties about the true value of these underlying frequentist probabilities. This line of thinking is common in the risk analysis community. The approach is referred to as the probability of frequency approach (Kaplan and Garrick, 1981). Conceptually, this approach makes sense provided that a meaningful population can be defined for the frequentist probabilities. In a Bayesian setting these probabilities are referred to as chances (Lindley, 2006; Singpurwalla, 2006). Möller *et al.* (2006) claim that the attitude of philosophers and statisticians towards second-order probabilities has been mostly negative, due to fears of an infinite regress of higher and higher orders of probability. This is, however, hard to understand, as noted also by Sahlin (1993), p. 26. Talking about uncertainties of a knowledge-based probability has no meaning.

However, the probabilities have limitations in capturing the relevant uncertainty aspects as discussed above. This also applies to the case of the probability of frequency approach. The knowledge-based probabilities are based on some background knowledge and this knowledge could be poor or wrong in many respects. How should we reflect this in the concept of risk and in our risk description? The answer is as in the above examples. We need to look beyond the probabilities. Probabilities alone would not fully capture the essence of the concept of risk. Uncertainty is a more fundamental concept than probability and should be the pillar of risk. We refer to Chapters 12 and 20. □

When a chance can be defined, there is some level of statistical stability – the outcome is 'predictable within limits', (Bergman, 2009). This was pointed out by Shewhart (1931, 1939) who introduced the concept of exchangeability – a sequence of random quantities is exchangeable if their joint probability distributions are independent of the order of the quantities in the sequence. Exchangeability is weaker than independence because, in general, exchangeable random quantities are dependent. It seems that Shewhart made this definition independently of de Finetti (1974) who used it to establish the so-called representation theorem. This theorem shows that chances exist as limits of exchangeable random quantities (Bernardo and Smith, 1994, p. 172). It justifies that chances can be viewed as underlying 'true probabilities' similarly to frequentist probabilities; see Singpurwalla (2006, p. 50).

Summary

It is common to define and describe risk using probabilities and probability distributions. However, these perspectives have been challenged. The probabilities could camouflage uncertainties (Rosa, 1998; Aven, 2009a, 2009b; Mosleh and Bier, 1996). The estimated or assigned probabilities are conditioned on a number of assumptions and suppositions. They depend on background knowledge. Uncertainties are often hidden in such background knowledge, and restricting attention to the estimated or assigned probabilities could camouflage factors that could produce surprising outcomes. By jumping directly into probabilities, important uncertainty aspects are easily truncated, meaning that potential surprises could be left unconsidered.

There are basically two ways of interpreting a probability:

1. A probability is interpreted as a relative frequency: the relative fraction of times the event would occur if the situation analysed were hypothetically 'repeated' an infinite number of times. The underlying probability is unknown and needs to be estimated. The estimation error is expressed by confidence intervals and the use of second-order knowledge-based probabilities.

2. Probability is a measure of uncertainty about future events and consequences, seen through the eyes of the assessor and based on some background information and knowledge. Probability is a knowledge-based (subjective) measure of uncertainty, conditional on the background knowledge (the Bayesian perspective).

References

Aven, T. (2003) *Foundations of Risk Analysis*, John Wiley & Sons, Ltd, Chichester.

Aven, T. (2008) *Risk Analysis. Assessing Uncertainties beyond Expected Values and Probabilities*, John Wiley & Sons, Ltd, Chichester.

Aven, T. (2009a) Uncertainty should replace probability in the definition of risk . Revised and resubmitted to *Reliability Engineering and System Safety*.

Aven, T. (2009b) Safety is the antonym of risk for some perspectives of risk. *Safety Science*, **47**, 925–930.

Aven, T. and Vinnem, J.E. (2007) *Risk Management, with Applications from the Offshore Oil and Gas Industry*, Springer-Verlag, New York.

Bergman, B. (2009) Conceptualistic pragmatism: a framework for Bayesian analysis? *IIE Transactions*, **41**, 86–93.

Bernardo, J.M. and Smith, A.F. (1994) *Bayesian Theory*, John Wiley & Sons, Inc., New York.

Borgonovo, E. and Apostolakis, G.E. (2001) A new importance measure in risk-informed decision making. *Reliability Engineering and System Safety,* **72**, 193–212.

Campbell, S. (2005) Determining overall risk. *Journal of Risk Research*, **8**, 569–581.

de Finetti, B. (1974) *Theory of Probability*, John Wiley & Sons, Inc., New York.

Flage, R., Aven, T. and Zio, E. (2009) Alternative representations of uncertainty in system reliability and risk analysis – review and discussion, in S. Martorell, C. Guedes Soares and J. Barnett (eds), *Safety, Reliability and Risk Analysis: Theory, Methods and Applications*, CRC Press, Boca Raton, FL.

Gärdenfors, P. and Sahlin, N.-E. (1988) Unreliable probabilities, risk taking, and decision making, in P. Gärdenfors and N.-E. Sahlin (eds), *Decision, Probability, and Utility*, Cambridge University Press, Cambridge, pp. 313–334.

Gillies, D. (2000) *Philosophical Theories of Probability*, Routledge, London.

Graham, J.D. and Weiner, J.B. (eds) (1995) *Risk versus Risk: Tradeoffs in Protecting Health and the Environment*, Harvard University Press, Cambridge, MA.

Haimes, Y.Y. (2004) *Risk Modeling, Assessment, and Management*, 2nd edn, John Wiley & Sons, Inc., Hoboken, NJ.

ISO (2002) Risk Management Vocabulary, ISO/IEC Guide 73.

Kaplan, S. (1991) Risk assessment and risk management – basic concepts and terminology, in R.A. Knief (ed.), *Risk Management: Expanding Horizons in Nuclear Power and Other Industries*, Hemisphere, New York, pp. 11–28.

Kaplan, S. and Garrick, B.J. (1981) On the quantitative definition of risk. *Risk Analysis*, **1**, 11–27.

Lindley, D.V. (2000) The philosophy of statistics. *The Statistician*, **49**, 293–337.

Lindley, D.V. (2006) *Understanding Uncertainty*, John Wiley & Sons, Inc., Hoboken, NJ.

Lowrance, W. (1976) *Of Acceptable Risk: Science and the Determination of Safety*, William Kaufmann Inc., Los Altos, CA.

Möller, N., Hansson, S.O. and Person, M. (2006) Safety is more than the antonym of risk. *Journal of Applied Philosophy*, **23**, 419–432.

Mosleh, A. and Bier, V.M. (1996) Uncertainty about probability: a reconciliation with the subjectivist viewpoint. *IEEE Transactions on Systems, Man and Cybernetics, Part A: Systems and Humans*, **26**, 303–310.

Polasek, W. (2000) The Bernoullis and the origin of probability theory: looking back after 300 years. *Resonance*, **5**, 26–42.

Rosa, E.A. (1998) Metatheoretical foundations for post-normal risk. *Journal of Risk Research*, **1**, 15–44.

Sahlin, N.-E. (1993) On higher order beliefs, in J.-P. Dubucs (ed.), *Philosophy of Probability*, Kluwer Academic Publishers, Dordrecht.

Savage, L.J. (1962) Subjective probability and statistical practice, in *The Foundations of Statistical Inference*, Methuen and John Wiley & Sons, Inc., London and New York.

Shewhart, W.A. (1931) *Economic Control of Quality of Manufactured Product*, Van Nostrand, New York.

Shewhart, W.A. (1939) *Statistical Method from the Viewpoint of Quality Control*, The Graduate School, Department of Agriculture, Washington, DC. Republished by Dover Publications, New York, with a new foreword by W.E. Deming in 1986.

Singpurwalla, N. (2006) *Reliability and Risk. A Bayesian Perspective*, John Wiley & Sons, Ltd, Chichester.

Vose, D. (2008) *Risk Analysis: A Quantitative Guide*, 3rd edn, John Wiley & Sons, Ltd, Chichester.

Further reading

Aven, T. (2009a) Uncertainty should replace probability in the definition of risk. Revised and resubmitted to *Reliability Engineering and System Safety*.

Aven, T. (2009b) Safety is the antonym of risk for some perspectives of risk. *Safety Science*, **47**, 925–930.

3

Risk equals a probability distribution quantile (value-at-risk)

In business, and in particular in the financial service industry, there has been an increasing focus on risk related to potential losses due to operational events such as fraudulent activity and rogue trading. In these contexts different types of risk measures are used. One of the most common is the value-at-risk (VaR) x_p, which equals the $100p\%$ quantile of the probability distribution of the potential loss X. Mathematically x_p is given by the formula $\mathrm{P}(X \leq x_p) = p$. Typical values of p are 0.99 and 0.999. VaR is the size of loss for which there is a small (e.g. 0.1%) probability of exceedance. Thus, if the VaR at probability level 99% is $100 million there is a only a 1% probability of a loss larger than $100 million. VaR is used in the determination of the amount of capital required to withstand adverse outcomes and avoid the enterprise becoming technically insolvent. VaR is also commonly used in investment analysis. Consider the example in Table 2.3. Here the VaR at level 95% equals a loss of $200 million. VaR captures the meaning of a worst case scenario for the investor. For example, in the case of Table 2.3, we are confident that the loss will not be larger than $200 million.

VaR has an intuitive appeal as a risk measure. However, it suffers from some problems, as will be discussed in the following.

Suppose that a bank's seven-day 99% VaR is $5 million. Then there is only a 1% probability that losses will exceed $5 million during a period of 7 days. One major problem with this risk measure is that it does not reflect the size of the potential losses exceeding $5 million. Consider the following example (Hull, 2006). Suppose a bank tells a trader that the one-day 99% VaR of the trader's portfolio must be kept at less than $10 million. Then there is a danger that the

Misconceptions of Risk T. Aven

The Basel Committee (2006) has developed a framework (the Basel II framework) for the determination of minimum capital requirements for banks and this refers to the use of VaR. It states:

> Given the continuing evolution of analytical approaches for operational risk, the Committee is not specifying the approach or distributional assumptions used to generate the operational risk measure for regulatory capital purposes. However, a bank must be able to demonstrate that its approach captures potentially severe 'tail' loss events. Whatever approach is used, a bank must demonstrate that its operational risk measure meets a soundness standard comparable to that of the internal ratings based approach for credit risk (i.e. comparable to a one year holding period and a 99.9 percent confidence interval).

trader will construct a portfolio where there is a 99% probability that the daily loss is less than \$10 million and a 1% probability that it is \$500 million. The trader is satisfying the risk limits imposed by the bank, but is probably taking unacceptable risks. Most traders would, of course, not behave in this way – but some might.

The problem is illustrated by Figure 3.1 which shows two probability distributions for the loss having the same VaR. However, the upper distribution in Figure 3.1 is much riskier than the lower distribution in the sense that the potential loss is much greater.

Hence, care has to be shown using VaR as a measure of risk. It is certainly necessary to look beyond VaR to provide a full description of risk.

Four criteria that a risk measure should satisfy

Artzner *et al.* (1999) have proposed four properties that a risk measure $r(X)(X =$ loss) should have, when using the measure to determine the capital required for withstanding adverse events. These are (Hull, 2006; Panjer, 2006):

1. **Monotonicity.** If an activity has lower losses than another activity for every state of the world, its risk measure should be lower; that is, if $X \leq Y$, then $r(X) \leq r(Y)$.
2. **Translation invariance.** If we add a fixed loss c to the loss of the activity, the risk measure should increase by c; that is, $r(X + c) = r(X) + c$.
3. **Homogeneity.** Changing the loss by a factor d should result in the risk measure being multiplied by d; that is, $r(dX) = dr(X)$.

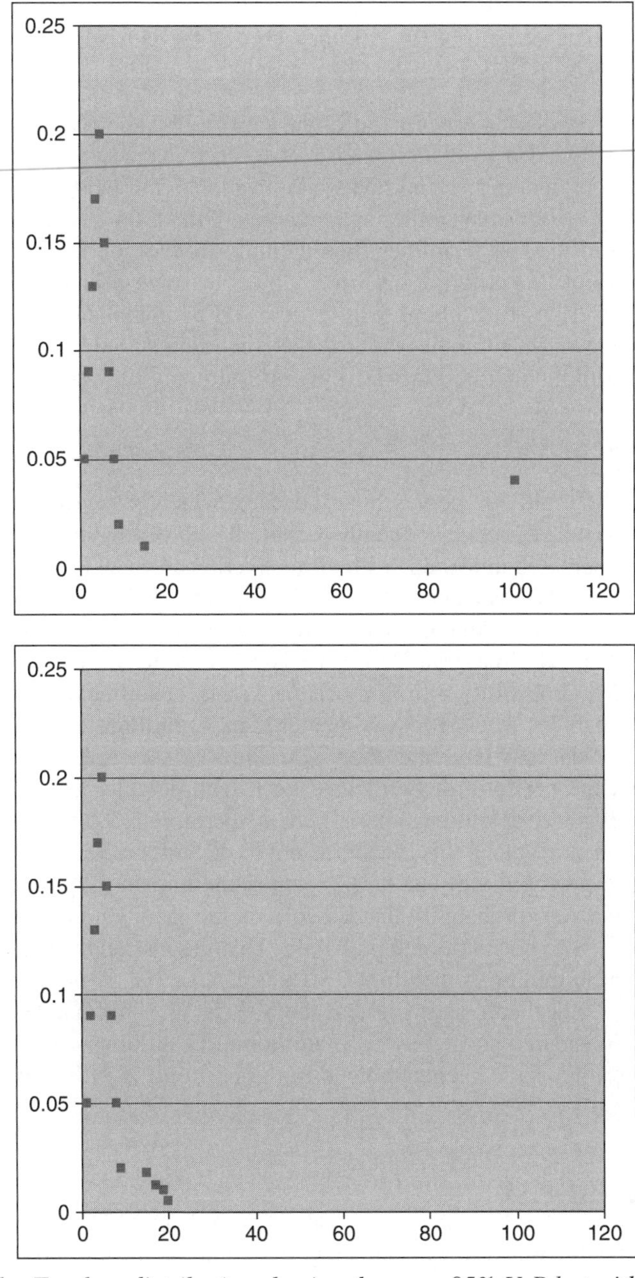

Figure 3.1 *Two loss distributions having the same 95% VaR but with completely different tails. VaR = 9 in both cases, and the maximum loss is 100 and 21, respectively.*

4. **Sub-additivity.** The risk measure for the sum of two losses should be no greater than the sum of the risk measures of the individual risk measures; that is, $r(X+Y) \leq r(X) + r(Y)$.

The first three conditions are straightforward given that the risk measure is the amount of cash needed to make the risk of the activity acceptable. The fourth condition reflects that there should be some diversification benefit from combining risks. When two loss quantities are aggregated, the total of the risk measures corresponding to the loss quantities should either decrease or stay the same. The appropriateness of this criterion has been subject to some debate. It is not obvious that the criterion should be met as there could be complicated dependencies between the losses. VaR satisfies the first three conditions, but it does not always satisfy the fourth, as will be illustrated by the following example.

Incoherence of VaR

Consider two \$10 million one-year loans, each of which has a 1.25% probability of defaulting (Hull, 2006). If a default occurs on one of the loans, the recovery of the loan principal is uncertain, with all recoveries between 0 and 100% being equally likely. If the loan does not default, a profit of \$200 000 is made. We suppose that if one loan defaults, it is certain that the other loan will not default. For a single loan, the one-year 99% VaR is \$2 million. To see this, note that there is a 1.25% probability of a loss occurring and, conditional on a loss, there is an 80% probability that the loss is greater than \$2 million. The unconditional probability that the loss is greater than \$2 million is 0.80 times 0.0125, or 1%. Consider next the portfolio of two loans. Each loan defaults with a probability 1.25% and they never default together. There is, therefore, a 2.5% probability that a default will occur. VaR in this case turns out to be \$5.8 million. This is because there is a 2.5% probability of one of the loans defaulting and, conditional on this event, there is a 40% probability that the loss on the loan that defaults is greater than \$6 million. The unconditional probability that the loss on the defaulting loan is greater than \$6 million is therefore 0.40×0.025, or 1%. A profit of \$200 000 is made on the other loan, showing that the VaR is \$5.8 million. The total VaR of the loans considered separately is \$2 million + \$2 million = \$4 million. The total VaR after they have been combined in the portfolio is \$1.8 million greater, that is, \$5.8 million.

Tail-value-at-risk

Alternative risk measures have been suggested, and one of the most commonly used is the tail-value-at-risk (TVaR). It is defined as the expected loss given that the loss exceeds the $100p\%$ quantile of the distribution (i.e. the VaR). If X denotes the loss, we may express TVaR at level p as

$$\text{TVaR} = \text{E}[X|X > x_p].$$

Alternatively we may write

$$\text{TVaR} = x_p + \text{E}[X - x_p | X > x_p],$$

where $\text{E}[X - x_p | X > x_p]$ is the expected excess loss. TVaR is also referred to as conditional tail expectation and expected shortfall (Panjer, 2006).

The TVaR measure is a consistent measure according to the four criteria stated above (Artzner *et al.*, 1999). Consider again the loan example (Hull, 2006). The VaR for a single loan is \$2 million. The expected shortfall from a single loan when the time horizon is 1 year and the probability level is 99% equals the expected loss on the loan, conditional on a loss greater than \$2 million. Given that losses are uniformly distributed, the expected value is halfway between \$2 million and \$10 million, that is, \$6 million (Hull, 2006). The VaR for a portfolio consisting of the two loans was calculated as \$5.8 million. The expected shortfall from the portfolio is, therefore, the expected loss on the portfolio, conditional on the loss being greater than \$5.8 million. The expected loss, given that we are in the part of the distribution between \$5.8 million and \$9.8 million, is \$7.8 million. This is therefore the TVaR for the portfolio. Because \$6 million + \$6 million > \$7.8 million, the TVaR satisfies the subadditivity condition for the example.

Reflection

In view of the discussion in Chapter 1, do you find that the concept of risk is captured by the TVaR value? Can decisions involving risk be based on TVaR?

For both questions, the answer is in general *no*. TVaR may be an informative risk measure but suffers from some of the same problems as identified for the expected value. It is easy to construct cases where the TVaR is equal for two cases but where the loss potential differs strongly. See the example in Table 3.1. The TVaR equals 5.0, but for case 2 there is a potential of a loss of 54 whereas case 1 has a maximum loss of 5. □

Table 3.1 Probability distributions for two cases with equal TVaRs (= 5.0) but different potential losses. The probability level is 95% and $x_p = 3$.

x	P(X = x), case 1	P(X = x), case 2
1	0.50	0.50
2	0.25	0.25
3	0.20	0.20
4	–	0.049
5	0.05	–
54	–	0.001

Computing VaR and TVaR

The above analysis requires probabilities and probability distributions to have been specified or estimated. Different techniques are used for this purpose, as noted for the probability distribution in Chapter 2:

1. Direct use of historical data
2. Direct assignments or estimates
3. Use of standard probability distribution, such as the Poisson distribution and the normal distribution
4. Use of detailed modelling of phenomena and processes, for example using event trees, fault trees and Bayesian belief networks. The analysis may use Monte Carlo simulation to obtain the probabilities based on the models.

Using the above definitions, risk is restricted to the probabilistic world. However, such a perspective can be challenged, as discussed in Chapter 2. A probability does not capture all aspects of concern, and this also applies to the VaR and TVaR measures. The problems apply very much to these measures since they relate to the tail of the probability distributions. It is extremely difficult to obtain rigorously developed, strongly supported tail distributions in most real-life situations. VaR and TVaR also have their limitations when all agree upon the probability distributions, as discussed above, but the specification/estimation problem is often the most critical one. We may compute historical-based VaR but its relevance may be poor due to rapid changes in industry and society in general. Thus, care has to be shown when using these risk measures.

Reflection

Is it essential that a risk measure meets the subadditivity criterion?

Some analysts think so (Szegö, 2004). They like to see risk being reduced when adding activities (returns, losses). Splitting up a company into different divisions (diversification) should reduce risk. However, it is obvious that cases of dependency exist where $X + Y$ could lead to a larger probability mass at extreme values than obtained by assuming independence. This does not mean that VaR is not an informative risk measure. The problem is not VaR. It simply reflects one aspect of the distribution, the quantile. It provides some information about the distribution, but it is should not be used alone to describe risk, as discussed above.

Summary

The VaR is a quantile of the loss distribution of interest. It is an informative risk measure but, of course, it does not capture the full information of the distribution and consequently it has to be used with care. The same applies to the

TVaR measure. Both measures are probability-based and the summary of the previous chapter is also relevant here: risk is more than computed probabilities and expected values.

References

Artzner, P., Delbaen, F., Eber, J.-M. and Heath, D. (1999) Coherent measures of risk. *Mathematical Finance*, **9** (3), 203–228.

Basel Committee (2006) *Basel Committee on Banking Supervision*, Bank for International Settlements, Basel

Hull, J. (2006) *Risk Management and Financial Institutions*, Prentice Hall, Upper Saddle River, NJ.

Panjer, H.H. (2006) *Operational Risk: Modeling Analytics*, John Wiley & Sons, Inc., Hoboken , NJ.

Szegö, G. (2004) Measures of risk. *European Journal of Operational Research*, **163**, 5–19.

Further reading

Szegö, G. (2004) Measures of risk. *European Journal of Operational Research*, **163**, 5–19.

4

Risk equals uncertainty

Risk is sometimes associated with uncertainty. For example, in Cabinet Office (2002) risk refers to uncertainty of outcome, actions and events. This perspective is most common in business contexts. This is a typical text found on the Internet:

> In the investment world risk equals uncertainty. Every investment carries some degree of risk because its returns are unpredictable. The degree of risk associated with a particular investment is known as its volatility.

The idea that risk equals uncertainty seems to be based on the assumption that the expected value is the point of reference and that it is known or fixed. Take, for example, the following problem of investing money in a stock market. Suppose the investor considers two alternatives, both with expectation 1, and variances 0.16 and 0.08, respectively. Since the expected value is the same for both alternatives, the focus would be on the variance, which is a measure of uncertainty relative to the expected value. Remember that the variance, $\text{Var}Y$, is defined by $\text{Var}Y = \text{E}(Y - \text{E}Y)^2$. As the second alternative has the lower risk (i.e. uncertainty), expressed by the variance, it would normally be chosen. Most people dislike uncertainties. Often standard deviation is used instead of the variance. The standard deviation of Y, $\text{SD}[Y]$, equals the square root of the variance of Y.

Securities that are volatile are considered risky because their return may change quickly in either direction. The standard deviation and variance express this risk by measuring the degree to which the securities fluctuate in relation to their expected return.

Portfolio analysis

Now suppose that you are considering investments in two stocks, (i) and (ii), with proportion w invested in stock (i) and $1 - w$ in stock (ii). Let Y_1 and Y_2 be

Misconceptions of Risk T. Aven

the returns for the two stocks and Y the total return, i.e. $Y = wY_1 + (1 - w)Y_2$. It follows that the expected return equals

$$\mathrm{E}[Y] = w\mathrm{E}[Y_1] + (1 - w)\mathrm{E}[Y_2],$$

and the variance is given by

$$\mathrm{Var}[Y] = w^2\ \mathrm{Var}[Y_1] + (1 - w)^2\ \mathrm{Var}[Y_2] + 2w(1 - w)\mathrm{Cov}(Y_1, Y_2), \qquad (4.1)$$

where $\mathrm{Cov}(Y_1, Y_2)$ is the covariance between Y_1 and Y_2. The covariance measures how much the two returns move in tandem. A positive covariance means that the returns move together, whereas a negative covariance means that they vary inversely (small returns in (i) correspond typically to large returns in (ii) and vice versa). Independence between Y_1 and Y_2 means a zero covariance.

To simplify, let us assume that the expected values for the stock returns both equal 1, $\mathrm{E}[Y_1] = \mathrm{E}[Y_2] = 1$, and the variance of the stock returns both equal 0.04, $\mathrm{Var}[Y_1] = \mathrm{Var}[Y_2] = 0.04$. Furthermore, assume that you consider only two alternatives:

A Invest in stock (i) only; that is, $w = 1$.

B Invest in both stocks equally; that is, $w = 1/2$.

It follows that $\mathrm{E}[Y] = 1$ for both strategies and the variance of Y when adopting strategy A equals 0.04. For strategy B, formula (4.1) gives

$$\mathrm{Var}[Y] = 0.01 + 0.01 + \mathrm{Cov}(Y_1, Y_2)/2 = 0.02 + 0.02\rho = 0.02(1 + \rho), \qquad (4.2)$$

where ρ is the correlation coefficient between Y_1 and Y_2, defined by

$$\rho = \frac{\mathrm{Cov}(Y_1, Y_2)}{\mathrm{SD}[Y_1]\ \mathrm{SD}[Y_2]}.$$

The correlation coefficient is a normalized version of the covariance. It takes values in $[-1, 1]$. Here $+1$ indicates perfect positive correlation and -1 perfect negative correlation.

We see from formula (4.2) that if the correlation coefficient ρ is zero, the return variance for strategy B is just the half of the variance in the case of strategy A. The uncertainties (risk) expressed by the variance are consequently smaller for strategy B than A. By mixing the two stocks we have reduced the risk (expressed by the variance). Normally that would lead to the conclusion that strategy B should be preferred to strategy A.

But why should we end diversification at only two stocks? If we diversify into a number of stocks (N) the volatility should continue to fall. However, we cannot reduce risk altogether since all stocks are affected by common macroeconomic factors (political events, war, etc.). The risk that remains after extensive diversification is called systematic risk (Bodie *et al.*, 2004). In contrast, the risk that can be eliminated by diversification is called unsystematic risk. We illustrate the difference between these two risk categories by extending the above example to N stocks. See the calculation box below.

The reference for the variance is the expected return. Another risk measure, the beta index (β), is based on a comparison with the market return. If the beta is close to 1, the stock's or fund's performance closely matches the market return. A beta greater than 1 indicates greater volatility than the overall market, whereas a beta less than 1 indicates less volatility than the market.

If, for example, a fund has a beta of 1.10 in relation to a specific market index, the fund has moved 10% more than the index. Hence, if the market index increased by 20%, the fund would be expected to increase by 22%.

The index beta is formally defined by

$$\beta = \frac{\mathrm{Cov}(Y, M)}{\mathrm{Var}(M)},$$

where Y is the fund return and M the market return.

Other measures used are the value-at-risk and tail-value-at-risk measures discussed in the previous chapter.

Variance for a Portfolio of N Stocks

Consider an investment in N different stocks, the same amount for each stock. Let Y_i be the return from stock i. Then the return of the total portfolio equals

$$Y = \frac{1}{N}\sum_{i=1}^{N} Y_i$$

The expected return is

$$\mathrm{E}Y = \mathrm{E}\left[\frac{1}{N}\sum_{i=1}^{N} Y_i\right] = \frac{1}{N}\sum_{i=1}^{N}\mathrm{E}Y_i,$$

and the variance equals

$$\mathrm{Var}Y = \mathrm{Var}\left(\frac{1}{N}\sum_{i=1}^{N} Y_i\right) = \sum_{i=1}^{N}\left(\frac{1}{N}\right)^2 \mathrm{Var}(Y_i) + \sum_{i=1}^{N}\sum_{j\neq i, j=1}^{N}\left(\frac{1}{N}\right)^2 \times \mathrm{Cov}(Y_i, Y_j) = \frac{1}{N}\overline{VAR} + \left(1 - \frac{1}{N}\right)\overline{COV} \quad (4.3)$$

where

$$\overline{VAR} = \frac{1}{N}\sum_{i=1}^{N}\mathrm{Var}(Y_i) \text{ and } \overline{COV} = \frac{1}{N^2 - N}\sum_{i=1}^{N}\sum_{j\neq i, j=1}^{N}\mathrm{Cov}(Y_i, Y_j).$$

Assuming that the variance $\text{Var}Y_i$ is bounded by a constant c, the average $\overline{VAR}$ is bounded by c, and hence the first term of (4.3) becomes negligible for large N. For large N, $\text{Var}Y$ approaches $\overline{COV}$. We refer to the terms of formula (4.3) as the unsystematic risk and the systematic risk, respectively. When the number of stocks is large, we see from (4.3) that the variance for the portfolio is approximately equal to the average covariance. The unsystematic risk is negligible when N is sufficiently large.

Reflection

The variance and standard deviation are risk measures which give the same weight to values above and below the expectation. Is that a reasonable property of a risk measure? Should not the focus be on the negative outcomes?

These measures are constructed to reveal variations around the expectation, and this is exactly what they do. They are also mathematically attractive. Other measures, such as quantiles of the probability distribution (VaR), can be used to identify possible extreme negative values.

Empirical counterparts

We must distinguish between the variance, standard deviation and beta assessed for a future return and observed values. We refer to the observed values as the empirical standard deviation, empirical variance and empirical beta. The empirical variance is denoted S^2 and is given by the formula

$$S^2 = \sum_i (Y_i - \overline{Y})^2/n,$$

where Y_i is the ith observation (return) and $\overline{Y}$ is the mean of the observations. Hence S denotes the empirical standard deviation.

The formula used to compute the beta (empirical beta) is based on the empirical covariance between the fund return and the market return, which equals

$$\sum_i (Y_i - \overline{Y})(M_i - \overline{M})/n.$$

The empirical beta β^* is equal to the empirical covariance normalized by the empirical variance of the market returns, that is

$$\beta^* = \frac{\sum_i (Y_i - \overline{Y})(M_i - \overline{M})}{\sum_i (M_i - \overline{M})^2}.$$

Table 4.1 and Figure 4.1 show an example of fund returns for 12 years together with the market returns. The means of the two data sets are 0.12 and 0.11,

Table 4.1 Returns for fund and market for 12 previous years.

Year	1	2	3	4	5	6	7	8	9	10	11	12
Market M	0.15	0.13	0.07	0.12	−0.04	0.31	0.23	0.31	0.02	−0.07	0.07	0.02
Fund Y	−0.05	0.05	0.01	0.25	0.04	0.15	0.40	0.29	0.33	−0.03	0.02	−0.02

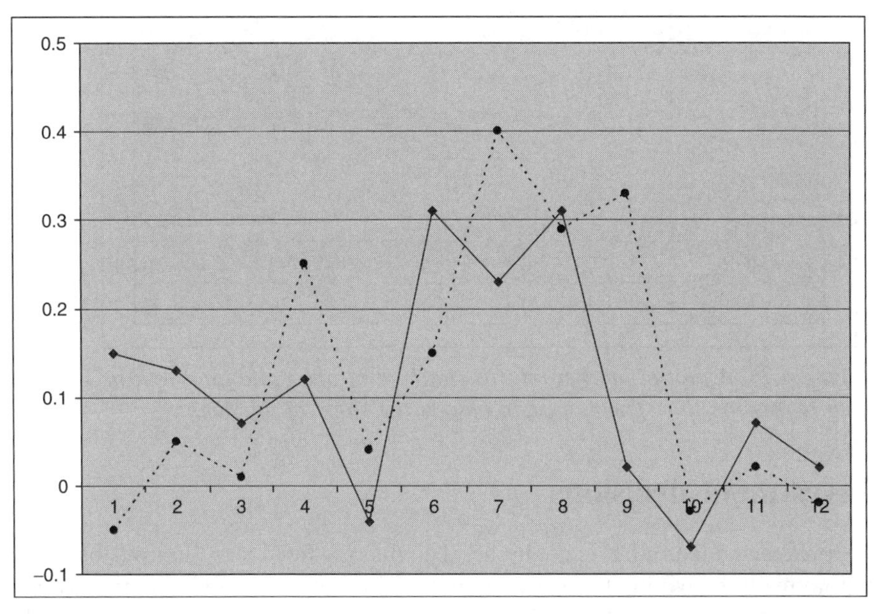

Figure 4.1 Returns for fund and market for 12 previous years (the fund returns are represented by the dotted line).

respectively, and the empirical standard deviations are 0.15 and 0.12, respectively. The empirical beta value β^* equals 0.63.

To find the empirical beta we can plot the market returns and fund returns in an m-y plot as shown in Figure 4.2. From this plot we identify the regression line $y = \alpha^* + \beta^* m$, where α^* and β^* are determined as the values of α and β that minimize the distance measure

$$\sum_i (Y_i - \alpha - \beta M_i)^2.$$

We find that

$$\alpha^* = \overline{Y} - \beta^* \overline{M}.$$

The α^* value is the point where the regression line intersects the y-axis. We see that in our case $\alpha^* = 0.05$. The formula for β^* is shown above; β^* represents the slope of the regression line.

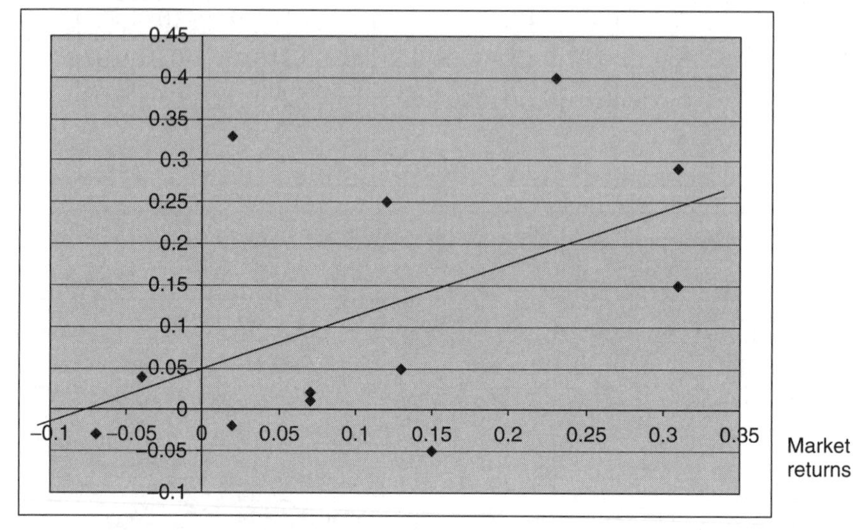

Figure 4.2 A plot of market returns (horizontal axis) and fund returns (y-axis), with regression line. The empirical beta is the slope of this line.

Investment decisions

If the expected return EY is not higher than the risk-free rate r, an investor would not normally be willing to invest any money in stocks. Investors are risk-averse. They need compensation for taking risk. We refer to the difference between the expected value and the risk-free rate as the risk premium. The difference between the actual return and the risk-free rate is called the excess return. Hence

$$\text{Risk premium} = \text{E[excess return]} = \mathrm{E}Y - r.$$

How large a risk premium should the investor require for investing in stocks? That depends on the risk (uncertainties), expressed, for example, by the standard deviation and beta index. The issue can be stated as a more general problem of obtaining an 'optimal' choice of securities, which has been thoroughly studied by economists and operations analysts for many years; see textbooks in finance such as Copeland *et al.* (2005), Weston and Shastri (2005) and Bodie *et al.* (2004). Many methods and models have been developed. It is, however, beyond the scope of this book to present and discuss these methods and models. We would, however, like to mention the capital asset price model (CAPM) as this model represents a theoretical *tour de force* (Fama and French, 2004). The main conclusion of the CAPM is that the price of the security will adjust to reflect the risk, so that its expected return is given by

$$\mathrm{E}[Y] = r + \beta(\mathrm{E}[M] - r), \tag{4.4}$$

where M is the market return. Equation (4.4) shows that the CAPM determines the expected return as the sum of the risk-free rate of return and β multiplied by the risk premium of the market $\mathrm{E}[M] - r$. The β parameter we may interpret as the number of systematic risk units. Thus the risk contribution is expressed by the risk premium of the market multiplied by the number of units of systematic risk. The attraction of the CAPM is that it is simple and offers powerful and pleasing return predictions. Its empirical performance is, however, being questioned. Some analysts conclude that it is, in fact, poor (Fama and French, 2004).

To verify model (4.4) we need a probabilistic basis. Let us first adopt the relative frequency interpretation of probability. The expected values in (4.4) then represent average returns in imaginary populations of an infinite number of similar securities and markets. These quantities (parameters) are unknown and must be estimated, leading to the following relationship between the corresponding estimators:

$$\mathrm{E}^*[Y] = r + \beta^*(\mathrm{E}^*[M] - r). \tag{4.5}$$

But it is obvious that $\mathrm{E}^*[Y]$ given by formula (4.5) could produce poor predictions of Y, the actual return for the security or securities. For one security it is obvious as the expected value is the average value, but also for a portfolio of diversified securities the predictions could be poor. Two main causes of the possible deviations are:

1. The model given by (4.5) is based on many assumptions, and these assumptions may, to varying degree, be correct for the case considered.

2. The estimates are poor compared to the underlying 'true' values. The data used to estimate the parameters may not be relevant for the future.

The first point is thoroughly discussed in the economic literature; see, for example Bodie *et al.* (2004) and Fama and French (2004). The second point relates to the use of historical data to predict the future and is the topic of Chapter 10. Modifications of the CAPM exist but any prediction of the return would obviously be subject to large uncertainties. If stock prices were predictable, what a gold mine that would have been. Part of the basis for the CAPM is the assumption of efficient markets, that is, markets in which all relevant information is reflected in the price of the security. However, real stock markets are not completely efficient, and other analysis tools, such as fundamental and technical analysis, are used to obtain information relevant to the future development of the value of the stocks. In fundamental analysis the focus is on the economic forces behind supply and demand that cause stock prices to increase, decrease or stay the same. In technical analysis, market actions are studied. Movements in the market are used to predict future changes in stock price.

If, alternatively, we adopt knowledge-based probabilities (subjective probabilities), the expected values in (4.4) represent centres of gravity in the uncertainty distributions assigned. Using (4.4) would then be a procedure for specifying $\mathrm{E}[Y]$

and again it is obvious that we may experience deviations between Y and $E[Y]$, even when we study a large portfolio. Equation (4.4) may not capture the key factors affecting the return and the data we use to establish the distribution may not be relevant for the future. Surprises relative to the history do occur all the time.

Expected value–variance analysis

Using the variance as a measure of uncertainty, one may question whether it would be sufficient to use the variance and the expected value as the basis for decision-making. This is discussed by Cox (2008) and Bodie *et al.* (2004), among others. Cox (2008) provides key references addressing the issue in the research literature and gives a simple illustrative example showing that the rational decision-maker would see beyond the variance and expected value. The main ideas of this example are summarized in the following.

Let X be the return for a possible investment, and let $\mu = EX$ and $\sigma^2 = \text{Var}X$. Now suppose that the investor bases his/her investment only on the expected value μ and variance σ^2, and that his/her preferences concerning what is an (un)acceptable investment are determined by an indifference curve as shown in Figure 4.3. Hence, an investment is considered acceptable if $\mu = 2$ and $\sigma^2 = 4$, but unacceptable if $\mu = 2$ and $\sigma^2 = 8$.

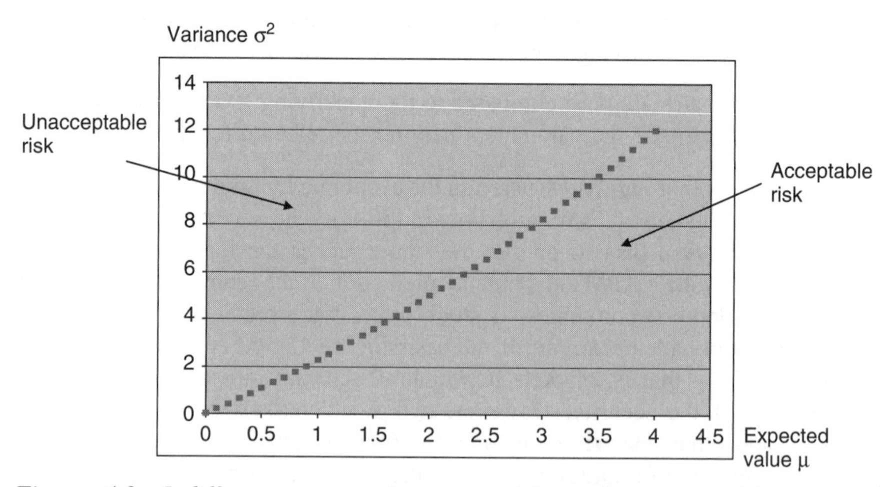

Figure 4.3 Indifference curve for acceptable and unacceptable expected value–variance relationship.

The slope of the indifference curve at the origin equals $s = 2$. Consider now a random variable Z that gives a positive return \$$2s$ with probability p and no return (\$0) with probability $1 - p$. We refer to Z as a Bernoulli random variable. It has an expected value equal to $2ps$ and a variance equal to $4s^2p(1 - p)$. As

p ranges from 0 to 1, the expected value–variance traces out a parabola with variance equal to 0 at $p = 0$ and at $p = 1$, and with maximum variance s^2 at $p = 0.5$ (see Figure 4.4). The slope of this curve at the origin is $2s$, which is seen by noting that a line from (0,0) to the point of the parabola corresponding to a particular value of p has slope $4s^2p(1-p)/2ps = 2s(1-p)$. As p approaches 0, this slope approaches $2s$. This means that the parabola starts above and to the left of the indifference curve through the origin, but it ends below and to the right of the indifference curve. Hence, the parabola must intersect the indifference curve somewhere above and to the right of the origin. Let p_0 be the value of p for this intersection point. The conclusion is that for $p < p_0$ the decision-maker finds the prospect Z unacceptable although it gives a positive win probability and no possibility of loss. Such a decision-maker is obviously not very rational.

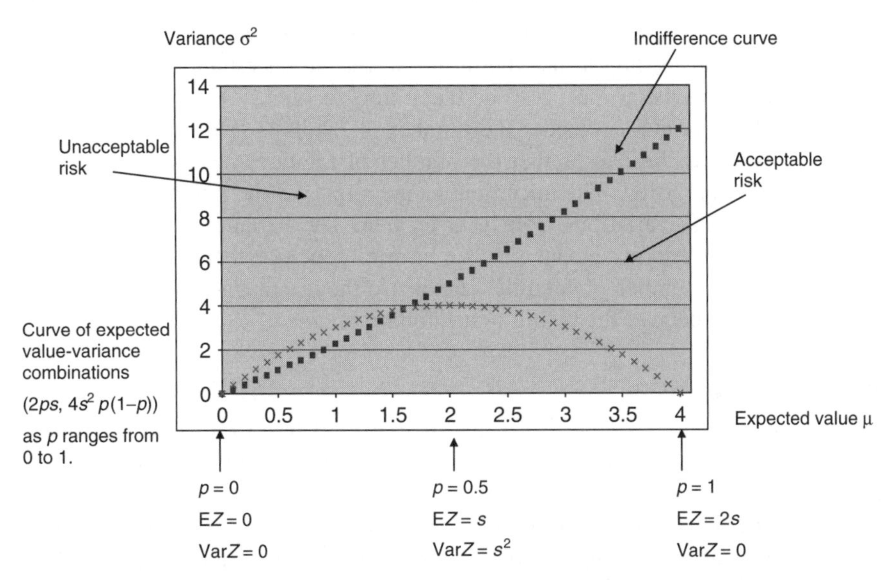

Figure 4.4 Indifference curve for acceptable and unacceptable expected value–variance relationship together with curve of expected value–variance combinations ($2ps$, $4s^2p(1-p)$) as p ranges from 0 to 1.

Thus it is not sufficient to just look at the expected value–variance relationship. The whole probability distribution must be considered. The previous two chapters also argued for looking beyond the probabilities.

Risk equals uncertainty as a general definition of risk

We have seen that risk is often viewed as uncertainty in financial contexts. Such a perspective is, however, problematical if we search for a general definition of risk. It fails to capture an essential aspect, the consequences dimension. Uncertainty cannot be isolated from the size, extension, and so on of the consequences. Take an extreme case where only two outcomes are possible, 0 and 1, corresponding respectively to no and one fatality, and the decision alternatives are A and B, having uncertainty (probability) distributions (0.5,0.5), and (0.0001, 0.9999), respectively. Hence, for alternative A there is a higher degree of uncertainty than for alternative B, meaning that risk according to this definition is higher for alternative A than for B. However, considering both dimensions, both uncertainty and the consequences, we would, of course, judge alternative B to have the highest risk as the negative outcome 1 is nearly certain to occur.

The expected values for the two alternatives are 0.5 and 0.9999, respectively, whereas the variances are 0.25 and 0.0001, respectively. This example is related to safety, but the conclusion is general and it also applies to investment problems. It does not make sense to say that there is negligible risk if we can predict with high confidence that the loss will be $10 billion. The uncertainties would be small, but it is misleading to use the term risk for the uncertainties as such. The consequences dimension must also be taken into account.

As another example, consider the number of fatalities in traffic next year in a specific country. We can predict the number of fatalities next year with quite a high level of precision. The uncertainties are slight, as the number of fatalities shows rather small variations from year to year. The variance is small. Hence, seeing risk as uncertainty means that we have to conclude that the risk is small, even though the number of fatalities are many thousands each year. Again we see that this perspective leads to a non-intuitive language.

Reflection

In the above example, suppose that the alternative B has a probability 1 for the outcome 1 to occur. Would you still say that the risk is greater for this case than for alternative A?

Some would say that the risk is zero in such a case as there are no uncertainties involved. But such reasoning is difficult to justify. If the probability of 0.9999 for the outcome 1 is increased to 0.99999, the risk should increase as it is more likely that the negative outcome occurs. We can continue increasing the probability and we should come to the same conclusion: the risk increases. Going to the limit 1 should not change this. The point is that risk always has to take into account the consequences and their severity.

Summary

Uncertainty seen in isolation from the consequences and the severity of the consequences cannot be used as a general definition of risk. Large uncertainties need attention only if the potential outcomes are large/severe in some respect.

In investment analysis, the uncertainty is seen in relation to a historical average value for similar investments. Risk captures the deviation and surprise dimension compared to this level. You may consider your investment to belong to a portfolio of projects which shows a historical value record of 5%. There are, however, factors that can cause the actual outcome to deviate from 5%, there are uncertainties about how large the deviation will be, and this is referred to as risk. The outcome X can thus be written

$$X = \mathrm{E}X + \text{risk (uncertainties)}.$$

This makes sense as long as EX is fixed or known. The number EX is just a reference number for comparisons. In the investment world, the focus is on portfolios and in such a context the uncertainties are referred to as systematic.

Probability-based measures are used to express the uncertainties (volatility) in this context, including the variance and standard deviation. As for all types of risk measures, they must be used with care as they do not capture the full information about risk.

References

Bodie, Z., Kane, A. and Marcus, A.J. (2004) *Investments*, 6th edn, McGraw-Hill, Boston.

Cabinet Office (2002) Risk: Improving Government's Capability to Handle Risk and Uncertainty, Strategy Unit report. UK.

Copeland, T.E., Weston, J.F. and Shastri, K. (2005) *Financial Theory and Corporate Policy*, Pearson, Boston.

Cox, L.A. (2008) Why risk is not variance: an expository note. *Risk Analysis*, **28**, 925–928.

Fama, E.F. and French, K.R. (2004) The capital asset pricing model: theory and evidence. *Journal of Economic Perspectives*, **18**, 25–46.

Further reading

Bodie, Z., Kane, A. and Marcus, A.J. (2004) *Investments*, 6th edn, McGraw-Hill, Boston.

Copeland, T.E., Weston, J.F. and Shastri, K. (2005) *Financial Theory and Corporate Policy*, Pearson, Boston.

5

Risk is equal to an event

In the social sciences, there are two prevailing definitions of risk:

(i) Risk is a situation or event where something of human value (including humans themselves) is at stake and where the outcome is uncertain (Rosa, 1998, 2003)

(ii) Risk is an uncertain consequence of an event or an activity with respect to something that humans value (IRGC, 2005).

These definitions are basically expressing the same thing: risk is an event or a consequence of an event (see Figure 5.1). The activity considered could produce events and consequences and these are subject to uncertainties. Something of human value is at stake.

The definitions express an *ontological realism* that specifies which states of the world are to be conceptualized as risk. As an objective state of the world, risk exists independently of our perceptions and knowledge of what is at risk and how likely a risk will be realized (Rosa, 1998, p. 28). Furthermore, by granting risk an ontological status, debates between risk paradigms are placed in an arena of disagreement over questions of knowledge, about out perceptions and understandings of risk, and about our understanding of how groups and societies choose to be concerned with some risks while ignoring others (Rosa, 1998).

Consider the activity of smoking. Humans have a stake in their health and well-being, and these values are threatened by events (such as uncontrolled growth of cells) that could lead to lung cancer and death. The events and consequences (outcomes) are subject to uncertainty. Or we could consider lung cancer as the 'event' in the definition and reformulate this as: humans have a stake in their health and well-being, and these values are threatened by lung cancer that could lead to weakened health and death. 'Lung cancer' is the risk according to definition (i). Risk is defined as 'an event (where . . .)'. 'Lung cancer' is the state

Misconceptions of Risk T. Aven

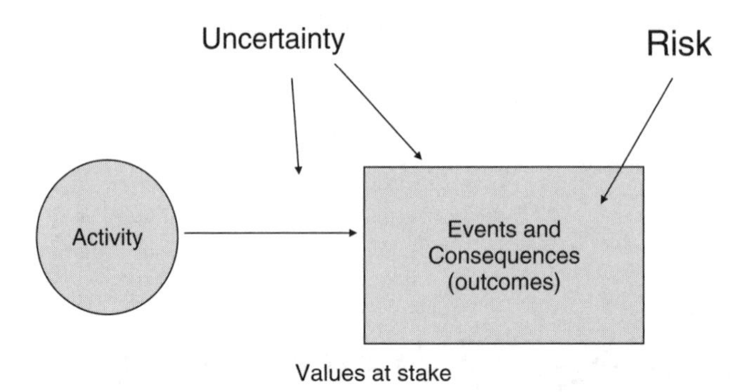

Figure 5.1 Risk defined as an event or a consequence (Aven and Renn, 2009a).

of the world. To cite Campbell (2006): 'Lung cancer is the risk you run if you smoke'. The event is subject to uncertainties and threatens humans' health and well-being.

An analogous definition to (ii) is found in reliability theory. Here the term unavailability is normally used as the expected fraction of time the system being considered is unavailable, that is, not functioning (Aven and Jensen, 1999), but we also see unavailability defined in sense (ii) as a state of the world, expressed by the actual fraction of time the system is unavailable (ISO, 2008). Then we may consider failures in the system as the 'events' according to definition (ii) and the fractional downtime as the consequences. The events and consequences are subject to uncertainties.

Risk, according to definitions (i) and (ii), has the following properties:

(a) It accommodates both undesirable and desirable outcomes.

(b) It addresses uncertainties instead of probabilities and expected values.

(c) It addresses outcome stakes instead of specific consequences.

These will be discussed in the following. We focus on definition (i).

Rosa (1998) makes a point of the fact that the definition can accommodate both undesirable and desirable outcomes. 'What humans value' is the key characteristic. In this way, economic risk as well as thrills are incorporated, since these domains of human risk action both address losses and gains. Accommodating both undesirable and desirable outcomes is a requirement for any definition of risk if we search for a widespread agreement. We will discuss this issue in more detail in Chapter 9.

Addressing uncertainties instead of probabilities and outcome stakes instead of specific consequences are crucial features of definition (i). They point to the challenge of establishing a precise measure of uncertainty as well as a precise measure of stakes involved. Probability is the common tool to express uncertainty, but the transformation of knowledge to probabilities is not straightforward.

The probability assignments are conditioned on a number of assumptions and suppositions. Uncertainties are often 'hidden' in probabilities and expected values, and restricting attention to these quantities could camouflage factors that could produce surprising outcomes. We refer to the discussion in Chapter 2.

In definition (i), Rosa (1998) refers to the two defining features of the risk: the 'outcome stake' and the 'uncertainty', and presents the illustration shown in Figure 5.2.

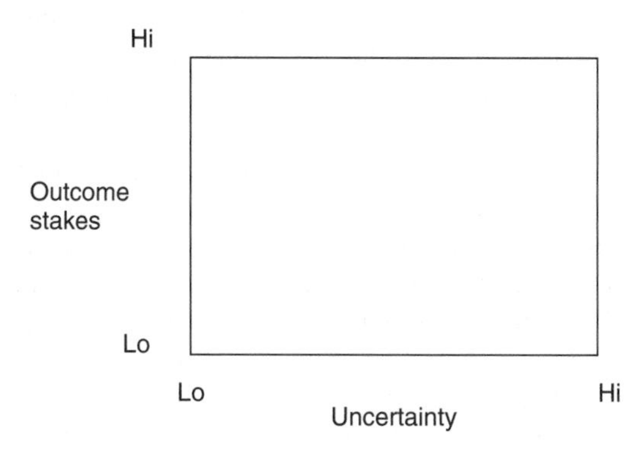

Figure 5.2 Dimensions of risk (Rosa, 1998).

Presenting the dimensions of risk in this way uncovers, according to Rosa, several important features of the risk concept. Firstly, the definition is consistent with the etymology of the word 'risk', meaning to navigate among dangerous rocks. Secondly, the representation is consistent with the standard definition of risk in the technical literature as probability times consequences (or outcomes).

A comment on this statement is in order. Definition (i) and Figure 5.2 relate to uncertainties, not probabilities. A probability is a measure of uncertainty, but uncertainties exist without specifying probabilities. Furthermore, there is nothing in definition (ii) and Figure 5.2 saying that we should multiply probabilities and consequences, that is, consider expected values. In the technical literature, risk is often defined by expected values, but as discussed in Chapter 2 it is equally common to define risk by the combination of probabilities and consequences (outcomes), that is, the full probability distribution of the outcomes (ISO, 2002; Aven, 2003).

Implications of seeing risk as an event or a consequence

Now let us examine how definition (i) is to be used in a risk research and/or a risk management context. Suppose you ask some experts on smoking 'Is the risk high?', 'How can risk be reduced?' and 'How is the risk compared to other risks?'. The experts would have to say that such statements have no meaning. They may consider the risk to be severe, but without an assessment of uncertainty or probability they cannot conclude whether the risk is high or low, and rank the risks.

Instead the experts must use statements like this: the (assigned or estimated) probability of the risk is high, and higher than the (assigned or estimated) probability of another risk. To reflect the existence of uncertainty without specifying probabilities, the experts may alternatively use a statement like this: the uncertainties about the occurrence of the risk are rather small, the uncertainties can be reduced by measures X and Y, and so on. We see that compared to standard terminology used in risk research and risk management, definition (i) leads to conceptual difficulties that are incompatible with the everyday use of risk in most applications.

As a more specific case, consider the ALARP principle, which is an important principle in risk management expressing the notion that risk should be reduced to a level that is as low as reasonably practicable (Aven and Vinnem, 2007). As it stands, this principle has no meaning according to definition (i), and has to be rephrased. 'Risk should be reduced' needs to be changed to 'Uncertainties/likelihoods of the risks should be reduced'. How this can be operationalized must be defined, but that is outside the scope of this book. The main point we would like to make here is that definition (i) cannot be applied to many risk concepts without making major detours and adding ad hoc assumptions.

As another example, we consider risk perception. Definition (i) means that risk and perception of risk are separate domains of the world (Rosa, 1998, p. 33). Risk exists independently of our perceptions. To get lung cancer is a risk (according to (i)), and how person A who smokes judges the risk is his/her risk perception. This judgement could, for example, be that he/she concludes that the risk is strongly undesirable, as the consequences of lung cancer are so severe. However, by looking at the likelihood (uncertainty) of the risk, he/she could find that this likelihood (uncertainty) is rather low and acceptable. The uncertainties/likelihoods need to be taken into account to reflect his/her perceptions. If we use the term risk perception according to (i), this is not an obvious interpretation, as risk is defined as 'the event lung cancer'. We again see that a detour and additional assumptions are required to apply a basic concept in risk research and risk management.

Reflection

To use definitions (i) and (ii), any judgement about risk needs to take into account uncertainties/likelihoods, so why not include this dimension into the risk concept?

It is a matter of taste, and the common language and standards in risk research and risk management do include the uncertainties/likelihoods in the risk concept. If considering risk an event of a consequence as in definitions (i) and (ii), the practice of risk research and risk assessment and management would need to be revamped. For example, we cannot talk about high risks and compare options with respect to risk.

Summary

Definitions (i) and (ii) mean that risk and the assessment of risk are separate domains of the world. The occurrence of a leakage in a process plant is a risk (according to (i)). This event is subject to uncertainties, but the risk concept is restricted to the event 'leakage' – the uncertainties and how people judge the uncertainties constitute a different domain. Hence, a risk assessment according to (i) and (ii) cannot conclude, for example, that the risk is high or low, or that option A has a lower or higher risk than option B, as it makes no sense to speak about a high or higher leakage. Instead, the assessment needs to conclude on the uncertainty or the probability of the risk being high or higher. A similar comment can be made on risk perception.

It is possible to build a risk framework for research and management based on definitions (i) and (ii), but compared to common terminology, they lead to conceptual difficulties that are incompatible with the everyday use of risk in most applications.

References

Aven, T. (2003) *Foundations of Risk Analysis: A Knowledge and Decision-oriented Perspective*, John Wiley & Sons, Ltd, Chichester.

Aven, T. and Jensen, U. (1999) *Stochastic Models in Reliability*, Springer-Verlag, New York.

Aven, T. and Renn, O. (2009a) On risk defined as an event where the outcome is uncertain. *Journal of Risk Research*, **12**, 1–11.

Aven, T. and Vinnem, J.E. (2007) *Risk Management, with Applications from the Offshore Oil and Gas Industry*, Springer-Verlag, New York.

Campbell, S. (2006) What is the Definition of Risk? Unpublished report.

IRGC (International Risk Governance Council) (2005) White Paper on Risk Governance. Towards an Integrative Approach, Author: O. Renn with Annexes by P. Graham. International Risk Governance Council, Geneva.

ISO (2002) *Risk Management Vocabulary*, ISO/IEC Guide 73.

ISO (2008) *Petroleum, Petrochemical and Natural Gas Industries: Production Assurance and Reliability Management*, ISO 20815:2008.

Rosa, E.A. (1998) Metatheoretical foundations for post-normal risk. *Journal of Risk Research*, **1**, 15–44.

Rosa, E.A. (2003) The logical structure of the social amplification of risk framework (SARF): metatheoretical foundation and policy implications, in N. Pidegeon, R.E. Kaspersen and P. Slovic (eds), *The Social Amplification of Risk*, Cambridge University Press, Cambridge.

Further reading

Aven, T. and Renn, O. (2009b) Response to Professor Eugene Rosa's viewpoint to our paper. *Journal of Risk Research*, in press.

Rosa, E. (2009) Viewpoint. The logical status of risk: to burnish or to dull? *Journal of Risk Research*, in press.

6

Risk equals expected disutility

As noted in Chapter 1, there is a need to look beyond the expected values in decision-making under uncertainty. Daniel Bernoulli (1700–1782) pointed this out more than 250 years ago: the decision must reflect the utility, referring to the level of goodness or badness of the outcomes. For example the harm of an accident needs to be taken into account as well as the severity of the economic loss.

But to what extent should the concept of risk cover the utility dimension? Some analysts and researchers conclude that it should do so completely (Campbell, 2005), and that the appropriate definition of risk is the expected utility, or the expected disutility if we prefer to focus on the harm (cost, damage, etc.) dimension. This means that the disutility of an outcome is weighted with its associated probability. Mathematically, we have

$$\text{Risk} = -\text{E}[u(C)],$$

where C is the outcome and $u(C)$ is the utility value for the outcome C.

The expected (dis)utility has strong theoretical support as it is the basis of some recognized theories of rational behaviour, including the von Neumann–Morgenstern expected utility theory and the subjective expected utility theory.

Reflection

Is it really important to discuss whether the risk concept should include the utility dimension or not?

Yes, as this is a discussion about more than concepts. It is about maintaining an important separation between assessment (science) and value judgements (management, politics). We refer to the following analysis.

Misconceptions of Risk T. Aven

Expected utility theory

The theory states that the decision alternative with the highest expected utility is the best alternative. As an example, let us again return to the Russian roulette type example of Chapter 2, where you win \$6000 if the die shows 1, 2, 3, 4 or 5, and you lose \$24 000 otherwise.

The possible outcomes are \$6000 and −\$24 000. Assuming the die is fair, the associated probabilities are 5/6 and 1/6, respectively. But without making such an assumption the probabilities need to be adjusted. We may, for example, use the same numbers as in Chapter 2, that is, 0.083 and 0.917 respectively, which are based on a probability of 90% that John is cheating. These probabilities are knowledge-based (subjective) probabilities.

The best outcome would obviously be \$6000, so let us give this consequence the utility value 1. The worst outcome would be −\$24 000, so let us give this outcome the utility value 0. It remains to assign a utility value to the outcome \$0 (which is the result if we do not play the game). Consider balls in an urn a proportion v of balls which are white. Let a ball be drawn at random; if the ball is white, the outcome \$6000 results, otherwise the consequence is −\$24 000. We refer to this lottery as '(6000, −24 000) with a chance of v'. How does '(6000, −24 000) with a chance of v' compare to achieving the result 0 with certainty? If $v = 1$ it is clearly better than 0, if $v = 0$ it is worse. If u increases, the gamble gets better. Hence there must be a value of v such that you are indifferent between '(6000, −24 000) with a chance of v' and a certain 0; call this number v_0. Were $v > v_0$, the urn gamble would improve and be better than 0; with $v < v_0$ it would be worse. This value v_0 is the utility value of the consequence 0. To be specific we may think of $v_0 = 0.70$, reflecting that we would not like to play if there is not a good chance of winning.

Hence we can compare the two alternative decisions: to play the game, where

$$\text{Expected utility} = \mathrm{E}u(C) = 0 \cdot \mathrm{P}(C = 0) + 1 \cdot \mathrm{P}(C = 1) = 0.083;$$

and not to play the game, where

$$\text{Expected utility} = \mathrm{E}u(C) = 0.70 \cdot 1 = 0.70.$$

The expected utility is therefore larger for the alternative not to play the game. The conclusion would change if we had used the fair die assumption as a basis, as the calculation then would be, for the decision to play the game,

$$\text{Expected utility} = \mathrm{E}u(C) = 0 \cdot \mathrm{P}(C = 0) + 1 \cdot \mathrm{P}(C = 1) = 5/6 = 0.833,$$

and for the decision not to play the game,

$$\text{Expected utility} = \mathrm{E}u(C) = 0.70 \times 1 = 0.70.$$

To change the conclusion we must have $v_0 > 5/6$.

The expected utility approach is attractive as it provides recommendations on a logical basis. If a person is coherent both in his preferences among consequences and in his opinions about uncertain quantities, it can be proved that the only sensible way for him/her to proceed is by maximizing expected utility. To be coherent when speaking about the assessment of uncertainties of events, a person must follow the rules of probability. When it comes to consequences, coherence means adherence to a set of axioms including the transitive axiom: if b is preferred to c, which is in turn preferred to d, then b is preferred to d. What we are doing is making an inference according to a principle of logic, namely that implication should be transitive. Given the framework in which such maximization is conducted, this approach provides a powerful tool for guiding decision-makers. Starting from such 'rational' conditions, it can be shown that this leads to the use of expected utility as the decision criterion. See, for example Savage (1972), von Neumann and Morgenstern (1944), Lindley (1985) and Bedford and Cooke (2001).

There are different versions of the expected utility theory, all inspired by Daniel Bernoulli's work. We have adopted Lindley's (1985) version which is based on knowledge-based probabilities. Von Neumann and Morgenstern (1944) use frequentist probabilities, which means that the theory is not applicable in most situations, for example the die example above. It is meaningless to speak about frequentist probabilities when assessing the event that John is cheating. Much of the theory in this area has been based on subjective probabilities using the betting interpretation, which means that uncertainty assessments are mixed with the desirability of rewards. In our view such a perspective is problematic in the same way as it is problematic that risk is defined through utilities. We will discuss this thoroughly in the following, but first some remarks on the weaknesses and limitations of the expected utility theory are in order.

The expected utility approach is established for an individual decision-maker. No coherent approach exists for making decisions by a group. K.J. Arrow proved in 1951 that it is impossible to establish a method for group decision-making which is both rational and democratic, based on four reasonable conditions that he felt should be fulfilled by a procedure for determining a group's preferences between a set of alternatives, as a function of the preferences of the group members. See Arrow (1951). A considerable body of literature has been spawned by Arrow's result, endeavouring to rescue the hope of creating satisfactory procedures for aggregating views in a group. But Arrow's result stands today as strong as ever. We refer to French and Ríos Insua (2000) and Watson and Buede (1987). Of course, if the group can reach a consensus on the judgements (i.e. the probabilities and utilities), then we are back to the single decision-maker situation. Unfortunately, life is not that simple in many cases – people have different views and preferences. Reaching a decision, then, is more about discourse and negotiations than mathematical optimization.

The expected utility theory is a normative theory stating how people should make their decisions. We know from research that people are not always rational in the sense that they adhere to the axioms of the theory; see the example below (the Allais paradox). A decision-maker would, in many cases, not seek to optimize and maximize his utility, but rather look for a course of action that is satisfactory. This idea, which is often referred to as a bounded rationality, is just one of many ways of characterizing how people make decisions in practice. Many theories have been developed to better reflect how people actually make their decisions. One of the most popular alternatives to the expected utility theory is the rank-dependent utility theory (Quiggin, 1981; Schmeidler, 1989; Tversky and Kahneman, 1992). Our concern in this book is risk concepts, whereas these theories face the challenge of providing structure for how people actually make decisions.

Example. The Allais paradox

Suppose you are offered a choice between two gambles:

1. \$0.5 million with probability of 1
2. \$2.5 million with probability 0.10, \$0.5 million with probability 0.89, and \$0 with probability 0.01.

Most people would prefer the first gamble, as they are guaranteed \$0.5 million. Suppose now that you are offered a choice between another two gambles:

3. \$0.5 million with probability 0.11, \$0 with probability 0.89
4. \$2.5 million with probability 0.10, \$0 with probability 0.90.

In this case many people would prefer gamble 4. This would not, however, be consistent with expected utility theory. From the first choice we have that

$$u(0.5) > 0.1 \cdot u(2.5) + 0.89 \cdot u(0.5) + 0.01 \cdot u(0),$$

and from the second choice we have that

$$0.1 \cdot u(2.5) + 0.9 \cdot u(0) > 0.11 \cdot u(0.5) + 0.89 \cdot u(0).$$

But these equations are inconsistent. By the first equation

$$0.11 \cdot u(0.5) - 0.01 \cdot u(0) > 0.1 \cdot u(2.5),$$

but by the second equation the inequality is reversed.

The example is based on Allais (1953). Most people do not like uncertainty and so prefer the better chance of winning \$0.5 million in betting option 1. In the latter case, the person favours gamble 4 because there is not much difference between 10% and 11%, but there is a big difference between \$0.5 million and \$2.5 million. This flies in the face of the so-called *independence axiom*, that rational choice between two alternatives should depend only on how those two alternatives differ. Yet, if the amounts involved in the problem are reduced to tens of dollars instead of millions of dollars, people's behaviour tends to fall back in line with the axioms of rational choice.

Should risk be separated from the utility dimension?

The situation addressed can be summarized as follows: an activity could produce events A, leading to consequences (outcomes) C, and the goodness/badness of these consequences is expressed by a utility function $u(C)$. The issue we now raise is whether the risk concept should apply to the A/C only or also relate to the utility of the consequences.

Risk relates to A/C only

The perspective means that risk is related to events and consequences of the world (fires, terrorist attacks, loss of lives, money, and so on), and that risk is separated from how we (the assessor, the decision-maker or other stakeholders) value the consequences (see Figure 6.1). In this way risk can be described by professional analysts and others without the risk characterization being mixed with difficult value judgements. Assigning a proper utility function is not easy (see the examples above and below), and whose utility function should be used? For many societal decision problems, there are a number of stakeholders and they may have completely different judgements about the importance of (say) a specific environmental damage. Some may also be reluctant to reveal their utility function as this would reduce their flexibility in negotiations and political processes. A decision-maker may prefer to base his/her decision on a professional characterization of uncertainties and likelihoods about events and consequences together with a more informal value judgement process. The distinction between professional risk analysis and the value judgement process is also reflected in

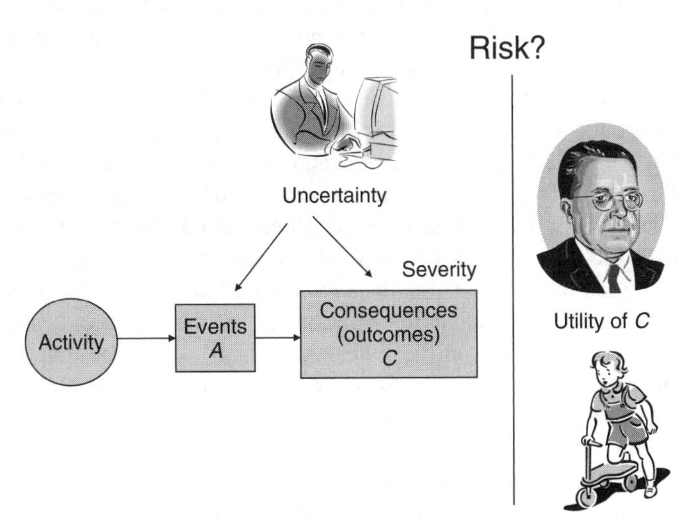

Figure 6.1 An illustration of the problem of restricting the risk concept to A and C and not incorporating the utility dimension.

common risk terminology (ISO, 2002, 2008): Risk analysis describes risk whereas risk evaluation broadens the picture to include aspects such as risk intolerability, cost–benefit balances, political priorities, potential for conflict resolution and social mobilization potential.

Risk also includes the utility dimension

A consequence of an accident or an investment could be loss of money, for example $1 million. The importance or severity of this loss may vary, however, depending on the wealth of those affected. This dimension can be reflected in the consequence classification system used for C, by introducing categories of severity, for example 'large losses' (suitably defined) and 'loss so severe that the company goes bankrupt'. This classification is still based on some objective measurable criteria. If we go on to ask what this consequence means for the relevant stakeholders, how bad or how good such a consequence is, we add a dimension of subjective utility. It would, in general, be impossible to agree about the utility function. Different persons value different goods differently. Nonetheless, we may add a utility function to the risk concept, making it clear whose utility function it is. Following this argument, the risk concept should include the utility dimension but not necessarily as expected disutility. Restricting risk to expected disutility means the adoption of one specific way of weighting the uncertainties and the utility function. The approach has a rationale, but it also suffers from some weaknesses as discussed above; see also the comments in the next section.

The consequence dimension needs to be included in the risk concept, but it may be discussed where this dimension ends. From a practical risk management point of view, it may be attractive to have a sharp line separating the risk analysis world, which characterizes the risk, and the value and management part which is concerned with the weight to be given to the risk relative to other concerns.

The need for a distinction between analysis (evidence) and values has been discussed thoroughly in the literature, in particular among social scientists. See, for example Renn (2008), Rosa (1998) and Shrader-Frechette (1991). To cite Renn (2008): 'The fact that evidence is never value-free and that values are never void of assumptions about evidence does not compromise the need for a functional distinction between the two'. Renn argues that in order to deal with risks one is forced to distinguish between what are likely to be the consequences when selecting option X rather than option Y, on the one hand, and what is more desirable or tolerable, on the other hand. Renn concludes that it is highly advisable to maintain the classic distinction between evidence and values and also to affirm that justifying claims for evidence versus values involves different routes of legitimization and validation.

Whether the risk concept includes or does not include, the utility dimension may not be considered essential as long as the distinction between analysis (evidence) and values is maintained. However, from a practical risk analysis and risk management point of view, the risk terminology should also be in line with this distinction, the consequence being that the risk concept and the risk description are associated with the left part of Figure 6.1 and we talk about risk evaluation, risk perception and risk handling when we incorporate the utility dimension.

Reflection

Can a risk analysis (assessment) be value-free?

No, an analysis is always contingent upon existing normative axioms and conventions. Some issues may be explored, others not, and this could significantly influence the outcome of the study. The analysis may be based on comparisons with established reference values for what constitute high or low figures. Again the result of the analyses could be strongly affected. And we could go on; obviously many aspects of the analysis are value-laden.

However, provided that the risk analysis simply makes predictions of the consequences of an activity and assesses associated uncertainties, the value dimension should not strongly influence the results. A good analysis should identify the aspects of the analysis for which values may affect the risk characterization.

Risk is more than expected disutility

If risk is defined as the expected disutility, we would experience the same type of problems as discussed in Chapter 2. We are conducting a probabilistic analysis based on a set of assumptions and these may hide uncertainties as shown by the die example where we assume that the die is fair. Any risk assessment would be based on a number of such assumptions, more or less identified. The conclusion is that risk is more than covered by the produced probabilities and expected values, including the expected disutility.

Finally, in this chapter we present an application of the expected utility theory when facing different attributes, benefit (money) and safety risk (fatalities as a result of accidents).

Example

In this example based on Aven (2003), we consider a decision problem with two alternatives: A and B. The possible consequences for alternative A and alternative B are $(2, X)$ and $(1, X)$, respectively. The first component of $(\cdot, \cdot)$ represents the benefit and X represents the number of fatalities, which is either 1 or 0. Assume the probabilities $\mathrm{P}(2, 0) = 0.95$, $\mathrm{P}(2, 1) = 0.05$, $\mathrm{P}(1, 0) = 0.99$ and $\mathrm{P}(1, 1) = 0.01$. The utility is a function of the consequences (i, X), $i = 1, 2$, and is denoted $u(i, X)$, and with values in the interval [0,1]. Hence we can write the expected utility

$$\mathrm{E}[u(i, X)] = u(i, 0)\mathrm{P}(X = 0) + u(i, 1)\mathrm{P}(X = 1).$$

To be able to compare the alternatives we need to specify the utility function u for the different outcomes (i, j). The standard procedure is to use the lottery approach explained above.

The best alternative would obviously be (2,0), so we give this consequence the utility value 1. The worst consequence would be (1,1), so we give this consequence the utility value 0. It remains to assign utility values to the consequences (2,1) and (1,0). Consider balls in an urn with v being the proportion of balls that are white. Let a ball be drawn at random; if the ball is white, the consequence (2,0) results, otherwise the consequence is (1,1). We refer to this lottery as '(2,0) with a chance of v'. How does '(2,0) with a chance of v' compare to achieving the consequences (1,0) with certainty? If $v = 1$ it is clearly better than (1,0), if $v = 0$ it is worse. If v increases, the gamble gets better. Hence there must be a value of v such that you are indifferent between '(2,0) with a chance of v' and a certain (1,0), call this number v_0. Were $v > v_0$, the urn gamble would improve and be better than (1,0); with $v < v_0$ it would be worse. This value v_0 is the utility value of the consequence (1,0). Similarly, we assign a value to (2,1), say v_1. As a numerical example we may think of $v_0 = 0.90$ and $v_1 = 0.10$, reflecting that we consider a life to have a higher value relative to the gain difference.

Based on this input we calculate the expected utility for the two alternatives. For alternative A,

$$1 \cdot \mathrm{P}(X = 0) + v_1 \cdot \mathrm{P}(X = 1) = 1.0 \cdot 0.95 + 0.1 \cdot 0.05 = 0.955,$$

while for alternative B,

$$v_0 \cdot \mathrm{P}(X = 0) + 0 \cdot \mathrm{P}(X = 1) = 0.9 \cdot 0.99 + 0 \cdot 0.01 = 0.891.$$

Risk is defined by the expected disutility, -0.955 and -0.891, respectively.

Now according to the utility-based approach, a decision maximizing the expected utility should be chosen. Thus alternative A is preferred to alternative B when the reference is the expected utility.

Alternatively to this approach, we could have specified utility functions $u_1(i)$ and $u_2(j)$ for the two attributes costs and fatalities respectively, such that

$$u(i, j) = k_1 u_1(i) + k_2 u_2(j), \tag{6.1}$$

where k_1 and k_2 are constants, with a sum equal to 1. We refer to Aven (2003, p. 125).

We see from this simple example that the task of assigning utility values for all possible outcomes could be challenging. The use of lotteries to produce the utilities is the adequate tool for performing trade-offs, but is hard to carry out in practice, in particular when there are many relevant factors, or attributes, measuring the performance of an alternative. To ease the specifications, several simplification procedures are presented. See, for example Bedford and Cooke (2001), Varian (1999) and Aven (2003). One possible approach is to define a parametric function for the utility function, and the value specification is reduced to assigning numbers to the parameters of this function. Examples of such procedures are found in Bedford and Cooke (2001), Varian (1999) and Aven (2003). This approach simplifies the specification process significantly, but it can be questioned whether the process imposes too strong a requirement on the specification of the utilities. Is the parametric function actually reflecting the decision-maker's preferences? The decision-maker should be sceptical about letting his preferences be specified more or less automatically without a careful reflection of what his/her preferences are. Complicated value judgements are not easily transformed into a mathematical formula. The specification of the utility function is particularly difficult when there are several attributes, and in most cases this is the case. For multi-attribute utility functions, simplifications can be performed by using weighted averages of the individual utility functions as illustrated by formula (6.1) (see also Clemen and Reilly, 2001). Again it is a question of whether the simplification can be justified. We refer to Bedford and Cooke (2001), Baron (2000) and Aven (2003).

In this example the probabilities are assigned and there is no reference to the background knowledge. As discussed above and in Chapter 2, the background knowledge may hide uncertainties and these should, as far as possible, be added to the calculated risk expressed through the probabilities and expected values.

Summary

If C is the outcomes (consequences) and $u(C)$ the utility function, risk defined by the expected disutility is given by $-\mathrm{E}u(C)$. Hence, the preferences of the decision-maker (and other stakeholders) are a part of the risk concept. The result is a mixture of professional assessments of uncertainties about C and the decision-makers' preferences concerning different values of C. This mixture is problematic. There will be a strong degree of arbitrariness in the choice of the utility function, and some decision-makers would also be reluctant to specify the utility function as it reduces their flexibility for weighting different concerns in specific cases. It should also be possible to describe risk in the case where the decision-maker is not able or willing to define his/her utility function.

References

Allais, M. (1953) Le comportement de l'homme rationnel devant le risque: critique des postulats et axiomes de l'école américaine. *Econometrica*, **21**, 503–546.

Arrow, K.J. (1951) *Social Choice and Individual Values*, John Wiley & Sons, Inc., New York.

Aven, T. (2003) *Foundations of Risk Analysis: A Knowledge and Decision-oriented Perspective*, John Wiley & Sons, Ltd, Chichester.

Baron, J. (2000) *Thinking and Deciding*, 3rd edn, Cambridge University Press, Cambridge.

Bedford, T. and Cooke, R. (2001) *Probabilistic Risk Analysis: Foundations and Methods*, Cambridge University Press, Cambridge.

Campbell, S. (2005) Determining overall risk. *Journal of Risk Research*, **8**, 569–581.

Clemen, R.T. and Reilly, T. (2001) *Making Hard Decisions with Decision Tools*, Duxbury/Thomson Learning, Pacific Grove, CA.

French, S. and Ríos Insua, D. (2000) *Statistical Decision Theory*, Arnold, London.

ISO (2002) *Risk Management Vocabulary*, ISO/IEC Guide 73.

ISO (2008) *Risk Management – General Guidelines for Principles and Implementation of Risk Management, Preliminary Version*.

Lindley, D.V. (1985) *Making Decisions*, John Wiley & Sons, Ltd, London.

Quiggin, J. (1981) Risk perception and risk aversion among Australian farmers. *Australian Journal of Agricultural Economics*, **25**, 160–169.

Renn, O. (2008) *Risk Governance. Coping with Uncertainty in a Complex World*, Earthscan, London.

Rosa, E.A. (1998) Metatheoretical foundations for post-normal risk. *Journal of Risk Research*, **1**, 15–44.

Savage, L.J. (1972) *The Foundations of Statistics*, Dover, New York.

Schmeidler, D. (1989) Subjective probability and expected utility without additivity. *Econometrica*, **57**, 571–587.

Shrader-Frechette, K.S. (1991) *Risk and Rationality: Philosophical Foundations for Populist Reforms*, University of California Press, Berkeley.

Tversky, A. and Kahneman, D. (1992) Advances in prospect theory: cumulative representation of uncertainty. *Journal of Risk and Uncertainty*, **5**, 297–323.

Varian, H.R. (1999) *Intermediate Microeconomics: A Modern Approach*, 5th edn, W.W. Norton, New York.

von Neumann, J. and Morgenstern, O. (1944) *Theory of Games and Economic Behaviour*, Princeton University Press, Princeton, NJ.

Watson, S.R. and Buede, D.M. (1987) *Decision Synthesis*, Cambridge University Press, New York.

7

Risk is restricted to the case of objective probabilities

In economic applications, a distinction has traditionally been made between risk and uncertainty, based on the availability of information. Under risk, the probability distribution of the performance measures can be assigned objectively, whereas under uncertainty these probabilities must be assigned or estimated on a subjective basis (Douglas, 1983). This definition goes back to Knight (1921). It is often referred to in the literature and is used by many researchers for structuring risk problems. See, for example Stirling *et al.* (2006). However, as we will see, other researchers and analysts find this definition quite meaningless. It is not compatible with most uses of the word 'risk' in today's society.

First let us go back Knight and his famous book from 1921. He writes (Knight, 1921, p. 233):

> To preserve the distinction which has been drawn in the last chapter between the measurable uncertainty and an unmeasurable one we may use the term 'risk' to designate the former and the term 'uncertainty' for the latter. The word 'risk' is ordinarily used in a loose way to refer to any sort of uncertainty viewed from the standpoint of the unfavorable contingency, and the term 'uncertainty' similarly with reference to the favorable outcome; we speak of the 'risk' of a loss, the 'uncertainty' of a gain. But if our reasoning so far is at all correct, there is a fatal ambiguity in these terms, which must be gotten rid of, and the use of the term 'risk' in connection with the measurable uncertainties or probabilities of insurance gives some justification for specializing the terms as just indicated. We can also employ the terms 'objective' and 'subjective' probability to designate

Misconceptions of Risk T. Aven

> the risk and uncertainty respectively, as these expressions are already in general use with a signification akin to that proposed.
>
> The practical difference between the two categories, risk and uncertainty, is that in the former the distribution of the outcome in a group of instances is known (either through calculation *a priori* or from statistics of past experience), while in the case of uncertainty this is not true, the reason being in general that it is impossible to form a group of instances, because the situation dealt with is in a high degree unique.

Knight refers to the concept of probability and considers three different types of probability situation (Knight, 1921, p. 194):

1. **A priori probability.** Absolutely homogeneous classification of instances completely identical except for really indeterminate factors. This judgement of probability is on the same logical plane as the propositions of mathematics (which also may be viewed, and are viewed by the writer, as 'ultimately' inductions from experience).

2. **Statistical probability.** Empirical evaluation of the frequency of association between predicates, not analysable into varying combinations of equally probable alternatives. It must be emphasized that any high degree of confidence that the proportions found in the past will hold in the future is still based on an *a priori* judgement of indeterminateness. Two complications are to be kept separate: first, the impossibility of eliminating all factors not really indeterminate; and, second, the impossibility of enumerating the equally probable alternatives involved and determining their mode of combination so as to evaluate the probability by *a priori* calculation. The main distinguishing characteristic of this type is that it rests on an empirical classification of instances.

3. **Estimates.** The distinction here is that there is *no valid basis of any kind* for classifying instances. This form of probability is involved in the greatest logical difficulties of all, and no very satisfactory discussion of it can be given, but its distinction from the other types must be emphasized and some of its complicated relations indicated.

In business the third category is most common. Knight (1921, p. 231) writes:

> The liability of opinion or estimate to error must be radically distinguished from probability or chance of either type, for there is no possibility of forming in any way groups of instances of sufficient homogeneity to make possible a quantitative determination of true probability. Business decisions, for example, deal with situations which are far too unique, generally speaking, for any sort of statistical tabulation to have any value for guidance. The conception of an

> objectively measurable probability or chance is simply inapplicable. The confusion arises from the fact that we do estimate the value or validity or dependability of our opinions and estimates, and such an estimate has the same form as a probability judgment; it is a ratio, expressed by a proper fraction. But in fact it appears to be meaningless and fatally misleading to speak of the probability, in an objective sense, that a judgment is correct. As there is little hope of breaking away from well-established linguistic usage, even when vicious, we propose to call the value of estimates a third type of probability judgment, insisting on its differences from the other types rather than its similarity to them.
>
> It is this third type of probability or uncertainty which has been neglected in economic theory, and which we propose to put in its rightful place.

According to Knight, risk is associated with the first two categories, whereas uncertainty is linked to the third. It seems that Knight considers business to mainly be concerning situations of uncertainty and not risk. Let us use some examples to investigate this definition in more detail. First we look at our die example.

Die example

John offers you a game: if the die shows a 6, you win $6000, otherwise you lose $24 000. Provided that the die is fair the probability distribution belongs to category 1 (a priori probability) as we can deduce the probabilities by logical arguments of symmetry. Hence, in this case the situation is characterized by risk.

As we do not know whether the die is fair, we need to perform a large number of trials to obtain an objective probability that the die shows 6: the relative frequency of 6s. This frequency produces the statistical probability of category 2. Again we classify the situation as one of risk.

However, this is a thought experiment as, in this case, we are not allowed to throw the die. The game consists of only one throw, without any testing of the die. The situation thus becomes one of category 3, a unique situation without a rigorous basis for specifying the probability. It is referred to as a situation of uncertainty and not risk.

A business example

A company considers the introduction of special equipment to improve production, and based on historical data from similar companies, an 'objective' distribution for the increased production can be generated. The costs can be accurately predicted and hence the distribution of the economic value – the net present value (see Chapter 18) – can be determined. Assuming that the company is large and has many such projects, the expected value should direct the decision as the expectation is a good prediction of the actual value. This is justified by the law of large numbers. We have established a situation of statistical probabilities and risk.

To be more specific, suppose that investment cost equals 0.1, and the income is either 0, 1 or 10, with probabilities 0.90, 0.09 and 0.01, respectively (to avoid technicalities we assume no *discounting* – this term is explained in Chapter 18). Then the expected economic value (benefit) is

$$-0.1 + 0 \cdot 0.90 + 1 \cdot 0.09 + 10 \cdot 0.01 = 0.09.$$

Hence, according to the expected value approach, the project should be implemented.

Now suppose that we are more or less ignorant about the consequences and outcomes of the project; the uncertainties are very large concerning the phenomena and processes involved. To be more specific, consider an example where the possible outcomes are -10^8, 0, 100 and 10^8. As a result of lack of information, probabilities of 0.25 are assigned to each possible outcome, which leads to an expected value equal to 25. Hence, if the expected value is used as a basis for the conclusion, the project should be accepted. However, it is obvious that the rationale for such a conclusion could be questioned. The expected value could produce an extremely poor prediction. Hence, we need to look beyond the expected value, and the probability distribution is presented to provide a more informative basis. However, the probabilities are not given much weight, as their basis is weak. To complicate the situation further, we allow for surprises to occur, resulting in even more extreme outcomes. A loss of 10^8 is severe, but the knowledge basis is poor and perhaps the losses could become even greater. A further complication is introduced if we also allow for outcomes of different dimensions: the losses could be economic but also safety-related. The possible consequences of the project or activity could be health problems for the workers in 20–30 years' time. The knowledge about the phenomena involved is not sufficiently understood to produce accurate predictions of the consequences.

The situation is obviously one in category 3 and uncertainty. Risk as a concept is not applicable when adopting Knight's definition.

Consider now a situation between the two extremes discussed above. It is characterized by some knowledge about similar types of systems and activities, but an 'objective' distribution does not exist. Probabilities of category 1 do not exist and statistical probabilities exist to some extent. We have some data but not sufficient to conclude that the numbers generated represent an objective distribution. Hence, the situation must be categorized as one of uncertainty and not risk.

Reflection

A huge insurance company sells fire insurance designed to indemnify the insured for loss of buildings as a result of fire. Seen from the insurance company's point of view, is the situation characterized by risk or uncertainty?

The probabilities can be classified as statistical probabilities and hence we may refer to a situation under risk.

Evaluation of Knight's work

For decades, economists and others have struggled to interpret Frank Knight's (1921) work. See, for example, discussions in LeRoy and Singell (1987), Holton (2004), Langlois and Cosgel (1993), Taylor (2003) and Runde (1998). Many interpretations of Knight's work have been presented and discussed. We will not go further into this discussion as it would give the wrong focus: trying to understand Knight's terminology. Remember that Knight did not have the same terminology and insights about the probability concept as we have today. The theory of subjective probability, for example, was not developed when Knight wrote his book. The basic idea, what is considered important in Knight's work in our context, is not the distinction between risk and uncertainty but the distinction between different decision problems or situations, one where it is possible to provide accurate predictions and one where this is not possible. Referring to the former situation as risk and the latter as uncertainty is unfortunate as it would exclude the risk concept from most situations of interest. In practice, we seldom have objective distributions, and should we adopt Knight's definition, we cannot refer to the concept of risk in these situations. If we adopt the subjective or Bayesian perspective on probability, Knight's definition of risk becomes empty. There are no objective probabilities. The terminology violates the intuitive interpretation of risk, which is related to situations of uncertainty and lack of predictability. In our view it is tragic that Knight's definition has been given the prominent place in the risk literature that it has. It has caused a lot of confusion, and many still refer to this definition as some sort of established terminology. Rather, we should leave the Knight nomenclature once and for all, but let us not forget his ideas about distinguishing between the different categories of decision problems. This idea has been an inspiration for many risk frameworks, as illustrated by the following examples.

Risk classification systems

Example 1

First we draw attention to the structure discussed by Aven and Flage (2009) and included in the above business example. Three types of decision situation are distinguished:

1. Known, 'objective' probability distributions can be established
2. More or less complete ignorance
3. A situation between the two extremes of 1 and 2.

The first category is characterized by populations comprising a large number of similar units, for example mass-produced technical items, individuals in a medical context or activities in a transport setting. The variation in the population constitutes the known 'objective' distribution. This variation is often referred to as aleatory uncertainty (Kaplan and Garrick, 1981) and corresponds to 'risk' in the Knightian framework defined above. The activity being studied is considered to be a unit of this population, and there are no uncertainties beyond the variation in the population.

The second category is characterized by more or less ignorance about the consequences and outcomes of the activity considered; refer to the business example above. Most situations in real life lie between these two extreme categories. In practice, the third category normally applies. We have some knowledge about similar types of systems and activities, but an 'objective' distribution does not exist.

The classification system is used in Aven and Flage (2009) as a framework for discussing the appropriate application of cost–benefit analyses (based on expected net present value calculations) and similar type of analyses to guide decision-making in a safety and production assurance setting. The cost–benefit analysis could play an important decision support role in situations in categories 1 and 3, but not so much in situations in category 2 (see also Chapter 18).

Example 2

The IRGC (2005) has introduced a classification system that also exhibits a resemblance with the Knight structure. The system is based on a categorization of risk problems according to complexity (linear to complex), uncertainty (high or low) and ambiguity (high or low) (Aven and Renn, 2009):

Linearity or simplicity is characterized by situations and problems with low complexity, uncertainty and ambiguity. Examples include car accidents, smoking, regularly reoccurring natural disasters, and safety devices for high buildings. Note that simplicity does not mean that the risks are low. The possible negative consequences could be very large. The point is that the cause–effect chain is linear and obvious for any observers, the uncertainties about the relationship between the event and the consequences are low and easy to determine and the values that are at stake are uncontroversial. It is possible to predict the occurrence of events and/or their consequences with a high degree of accuracy.

Complexity refers to the difficulty of identifying and quantifying causal links between a multitude of potential causal agents and specific observed effects. The nature of this difficulty may be traced back to interactive effects among these agents (synergism and antagonisms), long delay periods between cause and effect, inter-individual variation, intervening variables and others. Complex risk problems are often associated with major scientific dissent about dose–effect relationships or the alleged effectiveness of measures to decrease vulnerabilities. Examples of activities/systems with high complexity include sophisticated chemical facilities, synergistic effects of potentially toxic substances, failure of large interconnected infrastructures and critical loads to sensitive ecosystems.

Uncertainty refers to the difficulty of predicting the occurrence of events and/or their consequences based on incomplete or invalid databases, possible changes of the causal chains and their context conditions, extrapolation methods when making inferences from experimental results, modelling inaccuracies or variations in expert judgements. Uncertainty may result from an incomplete or inadequate reduction of complexity, and it often leads to expert dissent about the risk characterization. Examples of high uncertainty include many natural disasters (such as earthquakes), possible health effects of mass pollutants, acts of violence such as terrorism and sabotage, and long-term effects of introducing genetically modified species into the natural environment. For the risk of terrorism, the consequences of an attack can be fairly accurately predicted. However, the time and type of attack (event itself) is subject to large uncertainties.

The uncertainty in all these cases may be a result of 'known uncertainties' – we know what we do not know (e.g. that a known group of terrorists is planning a new attack, but we do not know where and when) – and 'unknown uncertainties' (ignorance or non-knowledge) – we do not know what we do not know (e.g. that a new terrorist group has been formed that attacks only special targets normally perceived as less attractive for the known terrorist groups).

Ambiguity refers to different views related to:

- the relevance, meaning and implications of the risk assessments for decision-making (interpretative ambiguity); or
- the values to be protected and the priorities to be made (normative ambiguity).

What does it mean, for example if neuronal activities in the human brain are intensified when subjects are exposed to electromagnetic radiation? Can this be interpreted as an adverse effect or is it just a bodily response without any health implications? Examples of high interpretative ambiguity include low-dose radiation (ionizing and non-ionizing), low concentrations of genotoxic substances, food supplements and hormone treatment of cattle. Normative ambiguities can be associated, for example with passive smoking, nuclear power, pre-natal genetic screening and genetically modified food.

The first category (linearity or simplicity) corresponds to the scope of the Knightian risk concept, but as we see, risk in the IRGC framework is not restricted to this category. Risk is present in all situations. The second and third categories (complexity and uncertainty) correspond to the scope of the Knightian uncertainty concept.

Table 7.1 shows the risk management implications of the risk problem categories. We refer to IRGC (2005) for details (see also Chapter 17). Here we are mainly concerned with the role of risk assessments. We see from Table 7.1 that risk assessments are considered a useful instrument within all four categories:

Table 7.1 Risk problem categorizations and their implications for risk management.

Risk problem category	Management strategy	Appropriate instruments	Stakeholder participation
Simple risk problem	Risk informed Routine-based risk treatment (risk reduction)	Statistical analysis Risk assessments Cost–benefit analyses Trial and error Technical standards Economic incentives Education, labelling, information Voluntary agreements	Not necessary to involve all stakeholders. *Instrumental discourse* among agency staff, directly affected groups as well as enforcement personnel is advisable.
Complexity induced risk problems	Risk informed (risk agent)	Risk assessments Cost–benefit analyses Characterizing the available evidence • Expert consensus seeking tools (e.g. Delphi) • Results fed into routine operation	Input for handling complexity could be provided by an *epistemological discourse* aimed at obtaining the best predictions of the occurrence of events and associated consequences. The goal is to resolve cognitive conflicts.
	Risk informed Robustness focused (risk absorbing system)	Risk assessments Cost–benefit analyses Improving buffer capacity of risk target through: • Additional safety factors • Redundancy and diversity in designing safety devices • Improving coping capacity • Establishing high-reliability organizations	

Table 7.1 (*continued*)

Risk problem category	Management strategy	Appropriate instruments	Stakeholder participation
Uncertainty induced risk problems	Risk informed and caution/precaution based (risk agent)	Risk assessments. Broad risk characterizations, highlighting uncertainties and features like persistence, ubiquity, and so on Tools include: Containment • ALARP (as low as reasonably practicable) • BACT (best available control technology), etc.	*Reflective discourse*: Include the main stakeholders in the evaluation process and search for consensus on the extra margin of safety in which they would be willing to invest in exchange for avoiding potentially catastrophic consequences. Deliberation relying on a collective reflection about balancing the possibilities for over- and under-protection.
	Risk informed Robustness and resilience focused (risk absorbing system)	Risk assessments. Broad risk characterizations Improving capability to cope with surprises: • Diversity of means to accomplish desired benefits • Avoiding high vulnerabilities • Allowing for flexible responses • Preparedness for adaptation	

(*continued overleaf*)

Table 7.1 (*continued*)

Risk problem category	Management strategy	Appropriate instruments	Stakeholder participation
Ambiguity induced risk problems	Risk informed and discourse based	Political processes Application of conflict resolution methods for reaching consensus or tolerance for risk evaluation results and management option selection: • Integration of stakeholder involvement in reaching closure • Emphasis on communication and social discourse	*Participative discourse*: competing arguments, beliefs and values are openly discussed.

Source: Aven and Renn (2009), modified from IRGC (2005), p. 47.

simple, complex, uncertain and ambiguous risk problems. The form and scope of the assessments, in particular the relevance for basing decisions on risk assessment results, would, however, differ strongly, depending on the problem at hand.

A Subjective Probability Interpretation of Frank Knight's work

An early generation of interpretations took the position that by risk Knight meant situations in which one could assign probabilities to outcomes and by uncertainty situations in which one could not (Langlois and Cosgel, 1993). As subjective probabilities can always be assigned, this type of interpretation made it possible to ignore situations of uncertainty entirely. All probabilistic situations are thus matters of risk.

However, such an interpretation is hard to justify in view of the quoted text above from Knight (1921). Rather it is reasonable to interpret Knight's definition as an attempt to distinguish between situations for which 'objective' probabilities can be determined, and where such probabilities cannot be determined.

References

Aven, T. and Flage, R. (2009) Use of decision criteria based on expected values to support decision-making in a production assurance and safety setting. *Reliability Engineering and System Safety*, **94**, 1491–1498.

Aven, T. and Renn, O. (2009) The role of quantitative risk assessments for characterizing risk and uncertainty and delineating appropriate risk management options, with special emphasis on terrorism risk. *Risk Analysis*, **29**, 587–600.

Douglas, E.J. (1983) *Managerial Economics: Theory, Practice and Problems*, 2nd edn, Prentice Hall, Englewood Cliffs, NJ.

Holton, G.A. (2004) Defining risk. *Financial Analysis Journal*, **60**, 19–25.

IRGC (International Risk Governance Council) (2005) Risk Governance – Towards an Integrative Approach, White Paper no 1, O. Renn with an Annex by P. Graham, IRGC, Geneva.

Kaplan, S. and Garrick, B.J. (1981) On the quantitative definition of risk. *Risk Analysis*, **1**, 11–27.

Knight, F.H. (1921) *Risk, Uncertainty and Profit*, Houghton Mifflin, Boston. Reprinted in 2002 by Beard Books, Washington, DC.

Langlois, R.N. and Cosgel, M. (1993) Frank Knight on risk, uncertainty, and the firm: a new interpretation. *Economic Inquiry*, **31**, 456–465.

LeRoy, S.F. and Singell, L.D. (1987) Knight on risk and uncertainty. *Journal of Political Economy*, **95**, 394–406.

Runde, J. (1998) Clarifying Frank Knight's discussion of the meaning of risk and uncertainty. *Cambridge Journal of Economics*, **22**, 539–546.

Stirling, A., Renn, O. and van Zwanenberg, P. (2006) A framework for the precautionary governance of food safety: integrating science and participation in the social appraisal of risk, in E. Fisher, J. Jones and R. von Schomberg (eds), *Implementing the Precautionary Principle*, Edward Elgar Publishing, Northampton.

Taylor, C.R. (2003) The role of risk versus the role of uncertainty in economic systems. *Agricultural Systems*, **75**, 251–264.

Further reading

Holton, G.A. (2004) Defining risk. *Financial Analysis Journal*, **60**, 19–25.

8

Risk is the same as risk perception

This chapter is about risk perception, that is, people's subjective judgement or appraisal of risk. Let us look at two examples.

Die example

Consider again the die example where John is offering you a game. If the die shows a 6, you lose $24 000, otherwise you win $6000. The associated probabilities are 1/6 and 5/6 provided that the die is fair. But you do not know whether the die you are throwing is a standard die or a special one designed for cheating you.

Your risk perception may be described as follows:

> John behaves strangely and you see no reason why he should offer you this game if he is not cheating. You consider the risk to be very high and too high for accepting the game.

Is this risk perception the same as risk and can it be seen as a risk description?

Industrial safety example

A chemical process plant is to be decommissioned. The plant is old, and the company that owns the plant would like to scrap and cover the plant *in situ*. People that live close to the plant, environmentalists and some of the political parties are sceptical about this plan. They fear pollution and damage to the environment. A large amount of chemicals has been used in the plant process. Their risk perception is that the uncertainties related to possible serious environmental damage and health problems are large and unacceptable. Certainly this is a judgement about

Misconceptions of Risk T. Aven

risk, but how is it linked to risk and the risk description provided by professional risk analysts? These issues will be further discussed below.

But first, let us examine in more depth what risk perception is and what influences the perception of risk. There is a huge body of literature on these issues, mainly the work of physiologists. Our brief review is based on Renn (1990, 2008) and Aven and Renn (2009, 2010).

Basic research about risk perception

Risk perception is a judgement (belief, appraisal) held by an individual, group or society about risk. There is no need for discussion about this definition, as long as the concept of risk is not defined. As we have seen, there are many ways of looking at risk, and consequently there are different interpretations of risk perception.

The risk perception may be influenced by the professional risk assessments and the individual's own risk assessment, as well as perceptional factors such as dread. Often the latter point is the most important one. Scientific assessments influence the individual response to risk only to the extent that they are integrated in the individual perceptions. Furthermore, relative frequencies are substituted by the strength of belief that people have about the occurrence of any undesirable effect. Both aspects are combined in a 'formula' that normally puts more weight on the magnitude of the effects than on the probability of their occurrence. The main insight is, however, that effects and likelihood are enriched by the perceived presence of situational and risk-specific characteristics that depend on properties such as the degree of perceived personal control, the perception of a social rather than an individual risk, or the familiarity of the risk situation (Slovic, 1987, 1992) (see Figure 8.1). Most cognitive psychologists believe that perceptions are formed by common-sense reasoning, personal experience, social communication and cultural traditions (Pidgeon, 1998). In relation to risk, it has been shown that humans link certain expectations, ideas, hopes, fears and emotions with activities or events that have uncertain consequences.

Some of the important elements influencing the perceptions of risks are (Renn and Rohrmann, 2000):

- intuitive heuristics and judgement processes with respect to probabilities and damages
- contextual factors relating to the perceived characteristics of the risk (e.g. familiarity or naturalness) and to the risk situation (e.g. voluntariness, personal controllability)
- semantic associations relating to the risk source, the people associated with the risk and the circumstances of the risk-taking situation
- trust and credibility of the actors involved in the risk debate.

Within the psychological domain of investigating risk perception, two major methodological approaches have been pursued: one through psychometric scaling

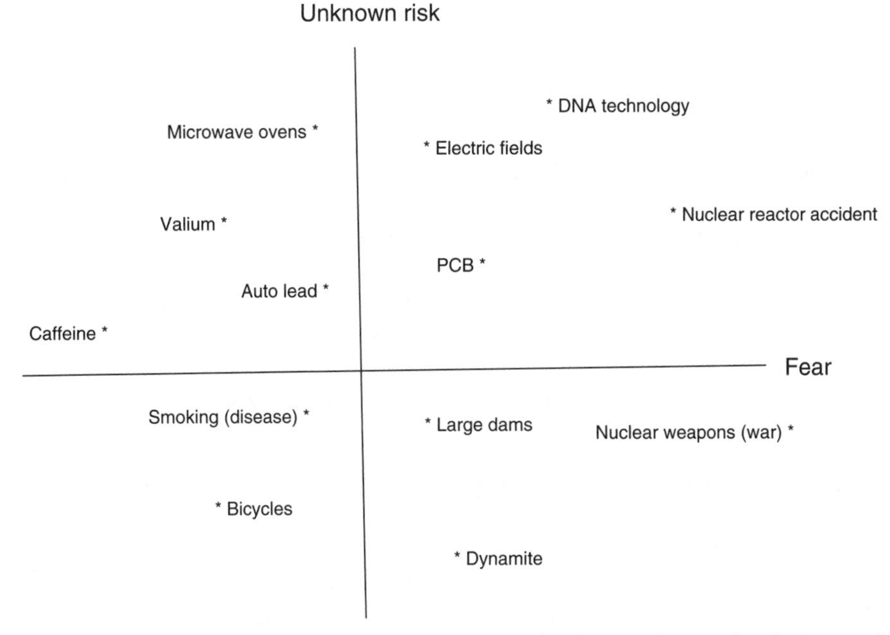

Figure 8.1 Risk perception seen in relation to the factors fear and unknown risk (the degree to which the risk is understood), expressed for a set of activities and technologies, based on a study of US students (Slovic, 1987, 1992). We see that the nuclear reactor accident score is high with respect to fear, whereas smoking scores relatively low for both dimensions.

exercises, by which individuals rate risks on attributes such as voluntariness, dread or familiarity, and the other one through mental models that reconstruct the mental associations of respondents between the risk and related subjects such as actors involved, context and attitudes towards the source of risk. Regardless of which of the two major methodological routes is pursued, there has been a clear consensus in the literature that the intuitive perception of risk refers to a multidimensional concept that cannot be reduced to the product of probabilities and consequences. Risk perceptions differ considerably among social and cultural groups. However, it appears to be a common characteristic in almost all countries in which perception studies have been performed, that most people form their judgements by referring to the nature of the risk, the cause of the risk, the associated benefits and the circumstances of risk-taking.

Risk perception studies can contribute to improving risk policies by:

- revealing public concerns and values
- serving as indicators for public preferences
- documenting desired lifestyles

- helping to design risk communication strategies
- representing personal experiences in ways that may not be available to the scientific assessment of risk. (Fischhoff, 1985)

In essence, psychological studies can help to create a more comprehensive set of decision options and to provide additional knowledge and normative criteria to evaluate them (Fischhoff, 1994).

There may be good reasons why evolution has provided human beings with a multidimensional, sophisticated concept of risk. This concept favours cautious approaches to new risks and induces little concern about risks to which everyone is already accustomed (Shrader-Frechette, 1991). It places relevance on aspects such as control and possibility for mitigation, both aspects that have been proven helpful in situations where predictions went wrong (Gigerenzer and Selten, 2001).

The difference between risk and risk perception

According to cultural theory and constructivism, *risk is the same as risk perception* (Jasanoff, 1999; critical comments in Rosa, 1998). Risk coincides with the perceptions of it (Douglas and Wildavsky, 1982; Freudenburg, 1989; Rayner, 1992; Wynne, 1992). Beck (1992, p. 55) concludes that 'because risks are risks in knowledge, perceptions of risks and risk are not different things, but one and the same'. Beck argues that the distinction between risk and risk perception is central to a scientific myth of expertise, according to which the population 'perceives risks' but science determines (i.e. identifies and quantifies) risk (Campbell and Currie, 2006, p. 152). A similar viewpoint can be found Jasanoff (1999, p. 150):

> I have suggested that the social sciences have deeply altered our understanding of what 'risk' means – from something real and physical if hard to measure, and accessible only to experts, to something constructed out of history and experience by experts and laypeople alike. Risk in this sense is culturally embedded in texture and meaning that vary from one social grouping to another. Trying to assess risk is therefore necessarily a social and political exercise, even when the methods employed are the seemingly technical routines of quantitative risk assessments . . . Environmental regulation calls for a more open-ended process, with multiple access points for dissenting views and unorthodox perspectives.

This viewpoint of risk being the same as risk perception is, however, not confined to these paradigms and scientists (Rosa, 1998). Rosa refers, for example, to the leading risk psychometrician Paul Slovic (1992, p. 119) who wrote: 'Human beings have invented the concept of "risk" . . . there is no such thing as "real risk" or "objective risk"'.

But rejecting the idea that there exists a 'real risk' or an 'objective risk', does not mean that risk is the same as perceived risk. Depending on how a probability is understood, risk definitions based on probabilities may or may not be associated with the idea of a 'real risk' or 'objective risk'. If probability is a way of expressing uncertainties, seen through the eyes of the assigner (a subjective or knowledge-based probability), there is no 'real risk' or 'objective risk'. However, subjective probabilities and related risk assignments are not the same as risk perception. The main difference is that risk perception is based on personal beliefs, affects and experiences irrespective of their validity. Subjective probabilities used in risk assessments are representations of individual and collective uncertainty assessments based on available statistical data, direct experience, models and theoretical approximations which all need justification that must also be plausible to others. The difference may be a judgement call at the borderline between subjective probabilities and perception, but the need for justification according to intersubjective rules is an important and relevant point of discrimination between the two concepts (Aven and Renn, 2009).

Furthermore, if we assume that risk perception does not only cover perceived seriousness of risk but also acceptability of risk, the difference becomes even more pronounced. Subjective probability carries no normative weight in terms of acceptability or tolerability of risk. You may assign a probability equal to 0.000000001 for an event to occur, but still find the risk to be intolerable. Our judgements about risk acceptability are, as we know from a number of risk perception studies, influenced by many factors outside the realm of probability. Perception of risk does not discriminate between the 'value-free' risk knowledge on the one hand and the value judgement about its acceptability or tolerability on the other hand. The assigned probability and judgements about risk tolerability or acceptability are different dimensions, or separate domains in the world of risk professionals who make a clear distinction between risk descriptions and judgements of acceptability.

In the case that the risk perspective is based on the idea that a true objective risk exists, it is obvious that the thesis 'risk = risk perception' is wrong (Campbell, 2005, p. 230). The above analysis showed, however, that this thesis is also invalid when probability is interpreted subjectively.

Let us return to the die and industrial safety examples. For the sake of simplicity we restrict attention to two risk perspectives:

(1) Risk is equal to the triplet (s_i, p_i, c_i), where s_i is the ith scenario, p_i is the probability of that scenario, and c_i is the consequence of the ith scenario (Kaplan and Garrick, 1981; Kaplan, 1991).

(2) Risk refers to uncertainty about and severity of the consequences (or outcomes) of an activity with respect to something that humans value (Aven and Renn, 2009).

In the first definition the probability is given either a relative frequency interpretation (1a) or a subjective (knowledge-based) interpretation (1b). These two definitions are general; the first is probability-based whereas the second is uncertainty-based.

Die example continued

First let us summarize what risk means in this case based on the two definitions (1) and (2):

- **Definition (1) of risk.** There are two scenarios: You win \$6000 or you lose \$24 000. The associated probabilities are p and $1 - p$, respectively. In case (1a) you estimate the probability p and perhaps you make a crude uncertainty assessment of p. In case (1b) you assign a subjective probability of 0.90 that the die shows 6 and you lose \$24 000.
- **Definition (2) of risk.** If the die is fair you lose \$24 000 with probability 1/6 and win \$6000 with probability 5/6. You have a strong belief that John is cheating. There are uncertainties about the fairness of the die.

In case (1a) risk is given by the pairs (−\$24 000, $1 - p$) and (\$6000, p), and we can conclude that risk perception is certainly not the same as risk. In case (1b) risk equals (−\$24 000, 0.90) and (\$6000, 0.10), and again we see that risk perception is different from risk. Risk in this case has no element of value judgements which capture our attitude to money and winning/losing money, as well as a judgement whether 0.90 is a high number or not.

For case (2), the conclusion is the same although the uncertainties are not quantified. As long as the risk concept and its description do not include any value judgements concerning attractiveness, desirability, and so on, risk and risk description are not the same as risk perception.

Industrial safety example continued

The company makes it clear that all chemicals will be removed. There should be no reason to fear possible health problems or environmental damage. However, several environmental organizations and the people who live in the neighbourhood of the plant are sceptical about the promises from the company. How can one be sure that all chemicals will be removed? They refer to the bad reputation this company has from similar activities internationally and the fact that it could be technically difficult to ensure that no 'surprises' occur in the future.

The company specifies detailed plans for the scrapping and conducts a risk assessment to demonstrate that the risk will be negligible. A consultant performs the assessment on behalf of the company. It produces risk results showing small probabilities of toxic chemicals. The conclusion is that the calculated risk is below what is normally considered negligible risk and hence acceptable. The chemicals do not represent a threat to the neighbours. The risk perspective is (1a).

The risk perceptions of the consultant and the company representatives are, to a large extent, governed by the risk assessment results. However, the risk description in the risk assessment includes no judgement about acceptability. Professional risk characterizations and risk acceptance are different domains. Nor does the risk picture cover any perceptional aspects such as fear. The number of people suffering from fear and so on could be seen as a consequence category of the activity in line with the number of deaths and injuries. However, the company would argue that this aspect should be given little attention as it is based on fear and not the scientific assessments.

For the neighbours, the risk perception extends far beyond the risk assessment results. The neighbours are not convinced by the risk numbers and they still perceive the risk as high and unacceptable. If risk refers to the calculated probabilities, risk perception is obviously a different domain also for the neighbours. For risk defined according to (2), this conclusion also holds as long as there is a clear distinction made between uncertainties and value judgements.

Summary

Risk perception is not the same as risk and professional risk characterizations, if we make a clear distinction between risk and uncertainty descriptions and value judgements. An event A and its consequences C are subject to uncertainties U, and we may use probabilities P to express these uncertainties, but that is a different domain from judgements about the extent to which we like or dislike A, C, U and P, which is the value judgment domain. Risk perception captures both the assessment aspect and the value judgement aspect. This is illustrated in Figure 8.2.

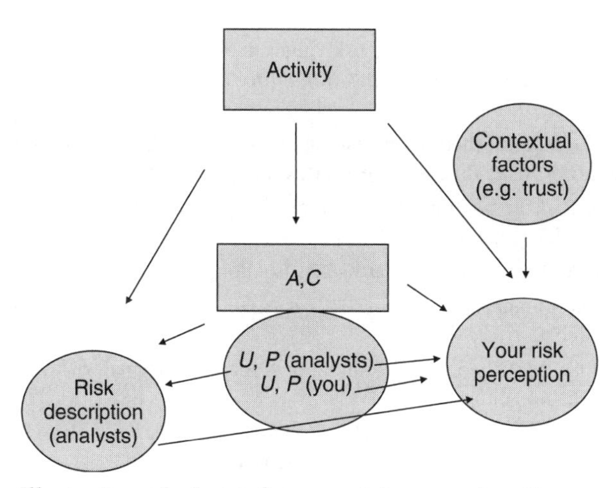

Figure 8.2 Illustration of what influences risk perception (A, events, C, consequences, U, uncertainty, P, probability).

References

Aven, T. and Renn, O. (2009) On risk defined as an event where the outcome is uncertain. *Journal of Risk Research*, **12**, 1–11.

Aven, T. and Renn, O. (2010) *Risk Management and Governance*, Springer-Verlag, New York, to appear.

Beck, U. (1992) *Risk Society: Toward a New Modernity* (translated by M.A. Ritter), Sage, London.

Campbell, S. (2005) Determining overall risk, *Journal of Risk Research*, **8**, 569–581.

Campbell, S. and Currie, G. (2006) Against Beck. In defence of risk analysis. *Philosophy of the Social Sciences*, **36**, 149–172.

Douglas, M. and Wildavsky, A. (1982) *Risk and Culture: The Selection of Technological and Environmental Dangers*, University of California Press, Berkeley.

Fischhoff, B. (1985) Managing risk perceptions. *Issues in Science and Technology*, **2**, 83–96.

Fischhoff, B. (1994) Acceptable risk: a conceptual proposal. *Risk: Health, Safety and Environment*, **1**, 1–28.

Freudenburg, W.R. (1989) Perceived risk, real risk: social science and the art of probabilistic risk assessment. *Science*, **242**, 44–49.

Gigerenzer, G. and Selten, R. (2001) Rethinking rationality, in G. Gigerenzer and R. Selten (eds), *Bounded Rationality. The Adaptive Toolbox*, MIT Press, Cambridge, MA.

Jasanoff, S. (1999) The songlines of risk. Environmental values. *Risk*, **8** (2) (Special Issue), 135–152.

Kaplan, S. (1991) Risk assessment and risk management – basic concepts and terminology, in R.A. Knief (ed.), *Risk Management: Expanding Horizons in Nuclear Power and Other Industries*, Hemisphere, New York, pp. 11–28.

Kaplan, S. and Garrick, B.J. (1981) On the quantitative definition of risk. *Risk Analysis*, **1**, 11–27.

Pidgeon, N.F. (1998) Risk assessment, risk values and the social science programme: why we do need risk perception research? *Reliability Engineering and System Safety*, **59**, 5–15.

Rayner, S. (1992) Cultural theory and risk analysis, in S. Krimsky and D. Golding (eds), *Social Theories of Risk*, Praeger, Westport, CT, pp. 83–115.

Renn, O. (1990) Risk perception and risk management: a review. *Risk Abstracts*, **7** (1), 1–9 (Part 1) and **7** (2), 1–9 (Part 2).

Renn, O. (2008) *Risk Governance*, Earthscan, London.

Renn, O. and Rohrmann, B. (eds) (2000) *Cross-cultural Risk Perception Research*, Kluwer, Dordrecht.

Rosa, E.A. (1998) Metatheoretical foundations for post-normal risk. *Journal of Risk Research*, **1**, 15–44.

Shrader-Frechette, K.S. (1991) *Risk and Rationality: Philosophical Foundations for Populist Reforms*, University of California Press, Berkeley.

Slovic, P. (1987) Perception of risk. *Science*, **236** (4799), 280–285.

Slovic, P. (1992) Perception of risk: reflections on the psychometric paradigm, in S. Krimsky and D. Golding (eds), *Social Theories of Risk*, Praeger, Westport, CT, pp. 153–178.

Wynne, B. (1992) Risk and social learning: reification to engagement, in S. Krimsky and D. Golding (eds), *Social Theories of Risk*, Praeger, Westport, CT, pp. 275–297.

Further reading

Okrent, D. and Pidgeon, N. (eds) (1998) Special issue on risk perception versus risk analysis, *Reliability Engineering and System Safety*. **59**, 1–159.

9

Risk relates to negative consequences only

In many applications risk relates to specific negative consequences, for example loss of life and injury. Common risk indices used are the expected number of fatalities and the expected number of injuries, suitably normalized (Aven, 2003). If you consult a dictionary, common definitions of risk are:

- exposure to the chance of injury or loss
- a hazard or dangerous chance
- the hazard or chance of loss
- the degree or probability of such loss.

And most people associate the word 'risk' with something undesirable and negative. For the positive outcomes we may introduce the term opportunity in a similar way as risk is linked to negative outcomes. Based on this separation between desirable and undesirable outcomes, we may define different risk perspectives. This distinction between risk and opportunity is common in, for example, project management. An example is given in Aven (2003):

> Risk index: the statistically expected value of the performance measure when only the possible negative outcomes are considered.

To illustrate this definition, consider the case where the performance measure C (representing, for example, income) can take four values, either $C = -5$, $C = -1$, $C = 1$ or $C = 2$, and the associated probabilities are 0.05, 0.20, 0.50, 0.25. Then, according to this definition,

$$\text{Risk} = -\text{E}[\text{minimum } \{0, C\}] = 5 \cdot 0.05 + 1 \cdot 0.20 = 0.45.$$

Misconceptions of Risk T. Aven

The possible positive outcomes are reflected in an opportunity index, which is defined as the statistically expected value of the performance measure when only the possible positive outcomes are considered. For the above numerical example, we obtain

$$\text{Opportunity} = \text{E}[\text{maximum}\ \{0, C\}] = 1.0.$$

The overall expected value of C, E[C], is equal to 0.55. □

The decision as to what is a positive or a negative outcome could be based on the expected value, if this value provides a meaningful reference level. In Chapter 4 the investment risk measures were related to uncertainties in relation to the expected value but did not distinguish between outcomes above or below the expected value. An obvious adjustment of these risk measures is to limit them to negative outcomes as compared to the expected values, and define analogous opportunity measures for positive outcomes.

Die example

In this example, John is offering you a game: if the die shows a 6, you lose $24 000, otherwise you win $6000. The associated probabilities are 1/6 and 5/6 provided that the die is fair. If risk is associated with the negative consequences, risk relates to the possible loss of $24 000, provided that the perspective is 'yours'. If we take John's perspective, risk is related to the possible loss of $6000. To support the decision whether to accept the offer or not, we have to cover all relevant consequences, both the negative and the positive, then assess uncertainties and assign probabilities. What name we give this assessment is not really important as long it is clear what we are trying to achieve. In this example it is easy to determine what is positive and what is negative, seen in relation to the relevant players; in other cases this is, however, not so straightforward. □

There are, however, good reasons for not restricting the risk concept to undesirable consequences, and many definitions of risk relate risk to both negative and positive consequences. What is a negative or undesirable consequence or outcome? To some, an outcome can be negative, and to others positive. One may wish to avoid a discussion on whether a consequence is classified in the correct category. Refer to the comment on this issue in Chapter 5, p. 56.

Investment example

A company is running a plant. A possible consequence of this activity is an accident leading to production loss. Does this event represent a risk if we adopt the risk–opportunity distinction? A production loss is negative for the company in a short-term perspective, but it may turn out that this break in production results in improvements to the plant which again lead to increased production and a reduced accident probability in the future (see Figure 9.1). So what is risk in this case and what is opportunity? We see that we will have a hard time trying to use these terms according to some strict rules. Alternatively, we may just use one term, and the natural candidate is risk.

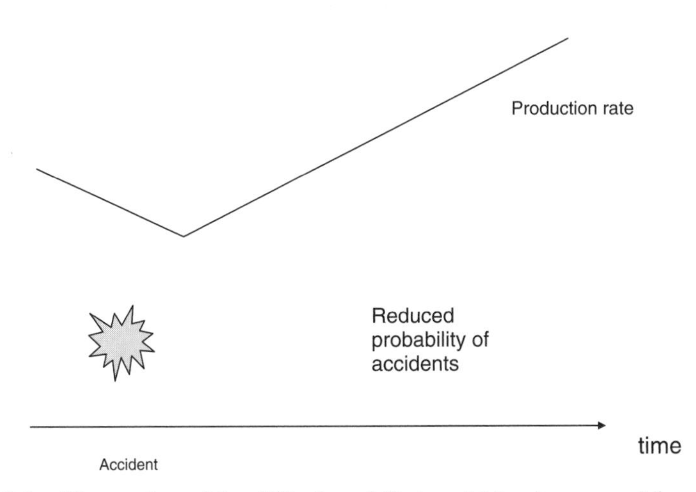

Figure 9.1 Illustration of the difficulty of distinguishing between risk and opportunity. Is the accident a negative consequence or a positive consequence? In the short term the accident leads to costs, but in the long term it results in higher production and reduced probability of accidents.

The word 'risk' derives from the early Italian *risicare*, which means 'to dare' (Bernstein, 1996). It was used by ancient sailors to warn the helmsman that rocks might be near. We need to dare to make decisions and act to achieve positive results, although the outcomes could turn out to be negative. We carry out an activity to obtain something positive, for example transporting a person from A to B, earning money, and so on, but something could go wrong. The rocks may, for example, cause the ship to sink. In this way one may say that a definition of risk which allows for both positive and negative consequences is consistent with the etymology of the word 'risk' (Rosa, 1998).

Summary

Most people associate the word risk with something undesirable and negative. However, restricting the concept of risk to negative outcomes only is problematic as it is often difficult to determine what is a negative outcome and what is a positive outcome. For example, in a short time perspective a failure event may be costly and undesirable, but in a longer perspective the failure may lead to changes that could lead to a higher level of performance.

References

Aven, T. (2003) *Foundations of Risk Analysis*, John Wiley & Sons, Ltd, Chichester.

Bernstein, P.L. (1996) *Against the Gods*, John Wiley & Sons, Inc., New York.

Rosa, E.A. (1998) Metatheoretical foundations for post-normal risk. *Journal of Risk Research*, **1**, 15–44.

10

Risk is determined by the historical data

To many people, risk is closely related to statistics. Numerous reports and tables are produced showing, for example, how the price of a certain product and the accident frequency vary over time. The statistics may report the price every day or as an average per month (or any other period). The prices of different products can be compared and averages of groups of products calculated. Analogously, accident data may be based on monthly or yearly reporting, and used to compare different activities. We refer to the following two examples.

Example. Product price

Table 10.1 and Figure 10.1 show the prices of two products over the last 12 months. Based on these data, we calculate the mean prices and empirical standard deviations:

Product	Mean ($\overline{Y}$)	Empirical standard deviation (S)
1	13.42	2.06
2	13.50	1.55

The empirical standard deviation is the square root of the empirical variance defined by

$$S^2 = \sum_i (Y_i - \overline{Y})^2/n$$

Misconceptions of Risk T. Aven

Table 10.1 Prices of two products in the last 12 months.

Month Product	1	2	3	4	5	6	7	8	9	10	11	12
1	10	11	13	12	13	12	14	13	14	15	17	17
2	12	11	12	14	14	14	16	15	16	13	12	13

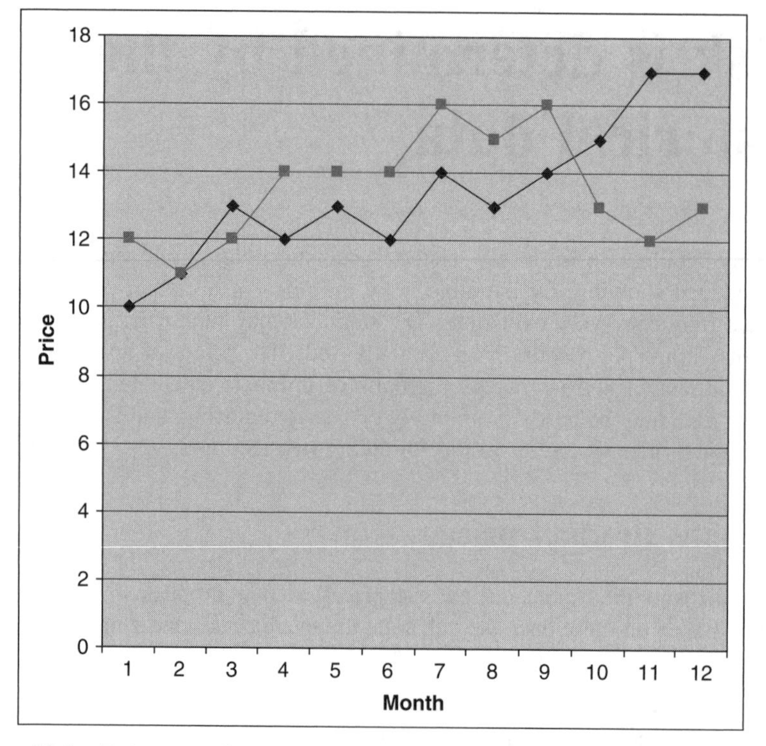

Figure 10.1 Price as a function of time for two products, based on the data in Table 10.1.

where Y_i is the price of the product in the ith month, $i = 1, 2, \ldots, 12$. We see that the mean prices for the two products are about the same, but the standard deviation is much larger for product 1. It seems that the price of product 1 has an increasing trend.

Example. Accident statistics

Figure 10.2 shows the number of serious injuries (properly defined) for an activity in a process plant based on 13 years of observations. These data can be split into subactivities, for example process and maintenance. The data in Figure 10.2 show that the injury rate is first increasing and then decreasing.

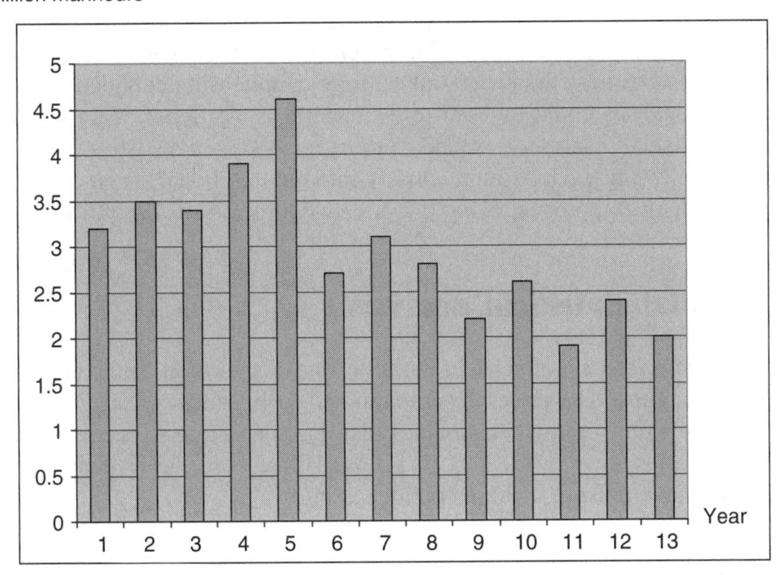

Figure 10.2 Normalized number of serious accidents in a process plant as a function of year.

Statistics and risk

Now the natural question to ask is: do these data provide information about the future, about risk?

Yes, although the data are historical data, they would usually provide a good picture of what to expect in the future. If the numbers of accidental deaths in traffic during the previous five years are 1000, 800, 700, 800, 750, we know a lot about risk, even though we have not explicitly expressed it by formulating predictions and uncertainties. The data indicate that we could expect about 700–800 fatalities next year. Looking at the data in Figure 10.2, about two serious injuries per million exposed hours are to be expected in the coming year. The conclusions for the price example are not so obvious. If we assumed no trend we would expect a price of about 13–14, but if we assume a trend for product 1 we would obtain a different prediction, about 17–18. But these numbers are predictions. They could turn out to be poor, relative to the actual numbers. This leads to uncertainty descriptions. How we carry out the uncertainty assessments depends on our risk perspective. We have to choose whether to adopt a traditional statistical approach or an alternative approach. If we adopt the traditional statistical approach we first need to define relevant parameters, for example the expected price next month (μ) or the probability that the price next month is below a certain number (p). This expected value and this probability are given a relative frequency interpretation. Hence μ is understood as the average price when considering an infinite number of similar months and p is the fraction of months where the price is not exceeding the specified number in this infinite population of similar months. Based on the observed data, we estimate the parameters and express the uncertainties, for example using confidence intervals. Regression analysis could be used to estimate possible trends. This would be a standard statistical analysis as found in most textbooks in statistics.

In the following we look more closely into the traditional statistical approach and an alternative approach, using the price example as an illustration.

Traditional statistical analysis

Let Y_i be the price of product 1 in month i, and $\overline{Y}$ the mean of the observations for the 12 months. The prices Y_i are assumed to be stochastically independent observations with a normal (Gaussian) distribution having expected value $\mu = \mathrm{E}[Y_i]$ and variance $\sigma^2 = \mathrm{Var}[Y_i]$. To estimate μ we use the estimator $\mu^* = \overline{Y}$. Uncertainty is expressed using a confidence interval.

As all Y_i are normally distributed and independent it follows that the mean $\overline{Y}$ is normally distributed and

$$V = \sqrt{n}(\overline{Y} - \mu)/S',$$

where $(S')^2 = \sum_i (Y_i - \overline{Y}^2)^2/(n-1)$, has a Student t distribution with $n-1$ degrees of freedom. In this distribution we know from statistical tables that the 95% quantile equals 1.80 and hence

$$\mathrm{P}(-1.80 \leq \sqrt{n}(\overline{Y} - \mu)/S' \leq 1.80) = 0.90.$$

From this we obtain

$$\mathrm{P}(\overline{Y} - 1.8 \cdot S'/\sqrt{11} \leq \mu \leq \overline{Y} + 1.8 \cdot S'/\sqrt{11}) = 0.90,$$

which gives a 90% confidence interval in our case equal to

$$(\overline{Y} - 1.8 \cdot S'/\sqrt{11}, \overline{Y} + 1.8 \cdot S'/\sqrt{11}).$$

For $\overline{Y} = 13.42$ and $S' = 2.15$, this interval is equal to (12.2, 14.6). Note that this is an uncertainty interval for μ, the average price, and not an uncertainty interval for a new observation.

The probability that the price for any arbitrary month exceeds, say, 15 is $P(Y_i > 15)$ which equals $1 - \Phi((15 - \mu)/\sigma)$, where Φ is the probability distribution function for the standard normal distribution with expected value 0 and variance 1. But μ and σ are unknown, so the best we can do is to estimate this probability. Replacing μ and σ with $\overline{Y}$ and S' respectively, we obtain

$$\text{Estimate of } P(Y_i > 15) = 1 - \Phi((15 - \overline{Y})/S') = 1 - 0.77 = 0.23.$$

We may also establish a confidence interval for the probability $p = P(Y_i > 15)$, using the fact that the random variables $Y_i - 15$ are independent and binomially distributed with success probability p. Standard procedures can then be used. We refer to textbooks in statistics. See also the example in Chapter 2, p. 21.

To perform a trend analysis in this setting, we assume a simple model; the expected price EY_i takes the linear form

$$EY_i = \alpha + i\beta.$$

Then standard regression analysis is used to estimate the parameters α and β, as shown in Chapter 4. Figure 10.3 shows the price data for product 1 and the estimated regression line. Based on the line, an estimated expected price for month 14 is 17.5. This number may also be seen as a prediction of the actual price at month 14.

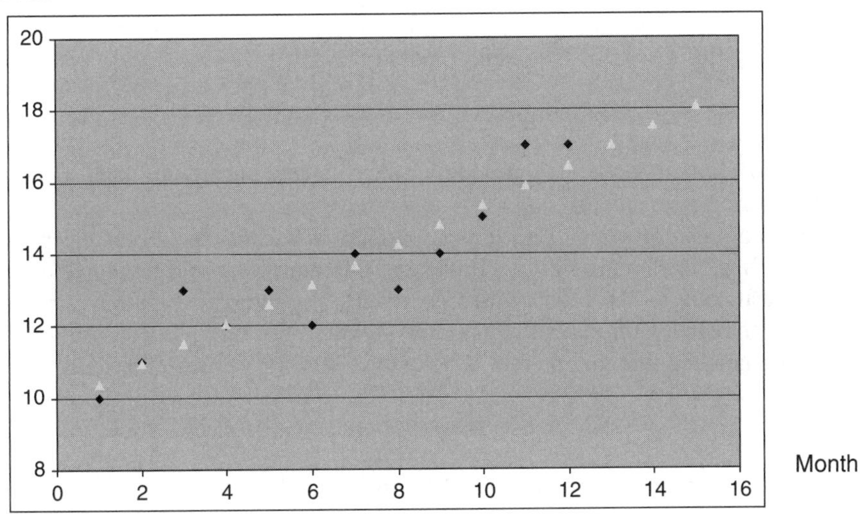

Figure 10.3 Price data for product 1 and the estimated regression line.

Confidence intervals can be established for the parameters α, β and $EY_i = \alpha + i\beta$. The slope of the line, β, is of special interest as it is a measure of the trend of the data. If $\beta = 0$ there is no trend and often the analysis is concerned with the extent to which the data prove that there is a trend present. Could the observed increase in the slope just be a result of 'randomness'? To estimate β we use the same estimator β^* as in Chapter 4:

$$\beta^* = \frac{\sum_i (Y_i - \overline{Y})(X_i - \overline{X})}{\sum_i (X_i - \overline{X})^2},$$

where $X_i = i$. The 90% confidence interval for β is given by

$$\beta^* \pm t_{n-2} \cdot S_\beta,$$

where t_{n-2} is the 95% quantile of the Student's t distribution with $n - 2$ degrees of freedom, and S_β is an estimator of the standard deviation of β^* given by

$$S_\beta^2 = \frac{\sum_i (Y_i - \alpha^* - i\beta^*)^2}{\sum_i (X_i - \overline{X})^2}.$$

We obtain the interval $0.55 \pm 1.81 \cdot 0.074 = (0.42, 0.68)$. Hence, the data provide strong evidence that the true β is positive.

Similarly, a confidence interval for $\mu_i = \alpha + i\beta$ can be formulated. It takes the form

$$\alpha^* + i\beta^* \pm t_{n-2} \left[\frac{\sum_j \left(Y_j - \alpha^* - j\beta^*\right)^2}{n-2} \right]^{1/2} \left[\frac{1}{n} + \frac{X_i - \overline{X}}{\sum_i \left(X_i - \overline{X}\right)^2} \right]^{1/2}.$$

Numerically this gives for the coming month $i = 13$:

$$17.0 \pm 1.81 \cdot 0.54 = (16.0, 18.0).$$

We can conclude that the data provide a rather narrow uncertainty band for the expected price next month.

One may also attempt to develop a prediction interval for the actual price next month, as in Berenson *et al.* (1988), but this interval would be a prediction interval estimate, as the probabilities generating the interval are unknown and must be estimated. Consequently, the interval would not serve its purpose as we have no guarantee that the interval is actually a 90% (say) prediction interval.

Discussion

The traditional statistical approach can be implemented when there are data available, as shown by the above example. To apply the approach, probability models like the normal distribution and the linear regression model need to be specified. The analysis is based on the assumption of a specific distribution and a type of form for the trend, which govern the future performance (randomness). But it is by no means obvious what the appropriate distribution and model should be. Who knows about the future? The historical data may indicate a certain distribution and trend, more or less excluding extreme observations, but this does not preclude such observations occurring in the future. Statistical analysis is based on the idea of a huge number of similar situations and if 'similar' is limited to the historical data, the population considered could be far too small. However, the statistician needs models to be able to perform the statistical analysis, and he/she will base his/her analysis on the data available. Taleb (2007) refers to the world of *mediocristan* and *extremistan* to explain the difference between the standard probability model context and the more extended population required to reflect extreme outcomes occurring in the future, respectively. Without explicitly formulating the thesis, Taleb (2007) says that we have to look beyond the historically-based probability models.

The statistical analysis above focuses on the parameters, for example $\mu_i = \alpha + i\beta$. A confidence interval is established for μ_i. But why be concerned about μ_i, the average price, when considering an infinite number of similar months to the one studied? What is the meaning of μ_i? Risk is about the possible occurrence of surprising – and, in particular, negative – prices in the future. The above analysis fails to reveal this in two ways:

1. The key quantity of interest is the actual price Y_i but the statistical analysis does not express uncertainties about Y_i. Instead, uncertainties are expressed about fictional parameters.

2. The analysis is based on the historical data and models explaining these, and gives little weight to possible extreme observations. The data may be more or less relevant for the future.

Some would respond to this that using the historical data is the best we can do, and the analysis must always be seen in relation to its assumptions. The analysis provides insights, although the assumptions may be invalid. Yes, an analysis must always be seen in the light of its assumptions, but to fully express risk we can and should do better than restrict attention to historical data and estimate fictional parameters.

An alternative approach

Let us perform a risk analysis based on the data we have observed and without introducing fictional parameters. Our focus is then the actual price in the coming months.

Table 10.1 and Figure 10.1 do not show the risks, they are just the historical numbers. However, the data may be used to say something about the future and risk. One way of doing this is outlined in the following. If we use the data as a basis, we obtain the following risk picture.

Let $Y(1)$ and $Y(2)$ be the prices for the products next month. Then, using the means as predictors, we obtain the following predictions for the Ys:

$$Y(1)^* = 13.4,$$
$$Y(2)^* = 13.5.$$

Furthermore, 90% prediction intervals can be specified based on knowledge-based probabilities (subjective probabilities):

$$\begin{aligned} Y(1) &: [11, 17], \\ Y(2) &: [12, 16]. \end{aligned} \tag{10.1}$$

These numbers are derived simply by requiring 90% of the observations to be in the interval.

If we use the normal distribution to express the probabilities, with expected values and standard deviation equal to the empirical quantities, we obtain the 90% prediction intervals

$$Y(1) : [10.0, 16.8],$$
$$Y(2) : [10.9, 16.1],$$

which are quite similar to the prediction intervals in (10.1).

We judge $(Y - \mathrm{E}Y)/S$ to be normally distributed with expectation 0 and variance 1. Let us look at the calculations for the first interval, $Y(1)$: [10.0, 16.8]. Since $(Y - \mathrm{E}Y)/S$ is normally distributed with expectation 0 and standard deviation 1, we obtain

$$\mathrm{P}(-c \leq (Y - \mathrm{E}Y)/S \leq c) = 0.90,$$

where $c = 1.645$. We find the quantile (c) from normal distribution tables. This yields

$$\mathrm{P}(\mathrm{E}Y - cS \leq Y \leq \mathrm{E}Y + cS) = 0.90,$$

and the interval $[\mathrm{E}Y - cS,\ \mathrm{E}Y + cS] = [10.0, 16.8]$ follows.

But, can we rely on these predictions? If there is a trend in price levels for product 1, it would be more reasonable to predict a price level of about 17 (next month) and not 13.4. If we use the same spread as above, we arrive at a prediction interval of [13, 21]. Only hindsight can show which is the best prediction, but the analysis makes it clear that a simple transformation of the historical figures can lead to very poor predictions.

By attempting to understand the data, by assuming a trend and carrying out a regression analysis, we may be able to improve the predictions. But we may also end up 'over-interpreting' the data in the sense that we look for all sorts of explanations for why the historical figures are as they are. Perhaps prices are rising; perhaps the trend arrow will be reversed next month. We can analyse possible underlying conditions that can affect prices, but it is not easy to reflect what the important factors are, and what is 'noise' or arbitrariness.

An analysis based on the historical numbers could easily become too narrow and imply that extreme outcomes are ignored, as discussed above. Surprises occur from time to time, and suddenly an event could occur that dramatically changes the development, with the consequence that the prices jump up or down. In a risk analysis such events should ideally be identified. However, the problem is that we do not always have the knowledge and insights to be able to identify such events, because they are extremely unexpected.

As a result, it is important to see the analysis in a larger context, where its constraints and boundaries are taken into account.

Reflection

For the price example, can you outline how to conduct the analysis when adopting the probability of frequency approach? (which is basically the same as a textbook Bayesian approach; see Chapter 15).

In this approach, the focus is on fictional parameters as in the traditional statistical approach, but subjective (knowledge-based) probabilities are used to assess uncertainties about these parameters. Suppose that our focus is on the expected price next month; $\mu_i = \alpha + i\beta$, where $i = 13$. Then this approach leads to second-order probabilities P expressing the analysts' uncertainty about μ_i. This analysis may, for example, result in a 90% credibility interval [10, 17] for the expected price next year, that is, $P(10 \leq \mu_i \leq 17) = 0.90$. The probability is a posterior distribution based on the 12-month observations. First a prior distribution for the parameters (α, β) is specified, and then the posterior distribution is computed using the standard Bayes updating procedure. See Chapter 15 and Ghosh *et al.* (2006) for details.

Conclusions

Historical data provide insights into risk. Assuming that the future will be as history shows, we may obtain good predictions about the future. However, there is in principle a huge step from history to risk as any assumption transforming the data to the future may be challenged. To fully express risk we need to look beyond historically-based data. Care should be exercised when using traditional statistical analysis to describe risk, as this analysis is based on strong assumptions. Risk is, to a large extent, about the aspects not included in these kinds of analysis, namely surprises. Sensitivity analysis is required to show how the results depend on key assumptions.

References

Berenson, M.L., Levine, D.M. and Rindskopf, D. (1988) *Applied Statistics. A First Course*, Prentice Hall, Englewood Cliffs, NJ.

Ghosh, J.K., Delampady, M. and Samanta, T. (2006) *An Introduction to Bayesian Analysis. Theory and Methods*, Springer-Verlag, New York.

Taleb, N.N. (2007) *The Black Swan: The Impact of the Highly Improbable*, Penguin, London.

Further reading

Taleb, N.N. (2007) *The Black Swan: The Impact of the Highly Improbable*, Penguin, London.

11

Risk assessments produce an objective risk picture

Risk assessments systematize the knowledge and uncertainties about the phenomena, processes, activities and systems being analysed. What are the possible hazards and threats, their causes and consequences? This knowledge and these uncertainties are described and discussed and this provides a basis for evaluating what is important (tolerable and acceptable) and for comparing options.

This is a way of describing the aim of risk assessment. The issue we discuss in this chapter is to what extent the risk picture produced by the risk assessment is objective in the sense that

(i) the results exist independently of the assessor, or

(ii) there is consensus among all stakeholders about the results.

Obviously, many examples exist where the risk assessments produce objective results according to these criteria. What is required for criterion (ii) is that all stakeholders agree upon the methods and procedures for describing risk. Consider the following examples.

Example. Standard statistical framework

Consider a very large (infinite) population of similar units and let p denote the fraction of units having a specific property A. Suppose n units are sampled from the population, and let X denote the number of units having property A.

This is an example of a standard statistical set-up, and well-established 'objective' methods exist for estimating p and producing confidence intervals for p reflecting variation in the data. We refer to the boxed example in Chapter 2 under the heading 'The meaning of a probability' and textbooks in statistics.

Misconceptions of Risk T. Aven

Example. Accidental deaths in traffic

Let us return to the accidental deaths in traffic example considered in Chapter 10. The numbers of accidental deaths in traffic during the previous five years are 1000, 800, 700, 800, 750. At a societal level these figures provide an informative risk characterization independent of the assessors and there is no dispute about the numbers.

Example. Risk level in Norwegian petroleum activities offshore

There was a dispute between the parties in the Norwegian petroleum sector in the late 1990s. Representatives of unions and authorities were concerned that the risk levels were increasing in offshore operations, whilst company management and their representatives claimed that 'safety had never been better'. Due to this dispute and some other factors, there was considerable mistrust between the parties and a lack of constructive communication on sensitive issues. There was a need for unbiased and, as far as possible, objective information about the actual conditions and developments. In order to meet these needs, the authorities, the Norwegian Petroleum Directorate, now the Petroleum Safety Authority Norway (PSA), set up a project ('the Risk Level Project') on extended indicators. The project is reported in Vinnem *et al.* (2006). The approach is based on recording the occurrence of near misses and relevant incidents, performance of barriers and results from technical studies. Of similar importance is an evaluation of the safety culture, motivation, communication and perceived risk. This is covered through the use of social science methods, such as questionnaire surveys and a number of interviews, audit and inspection reports as well as accident and incident investigations. A group of people with strong competence in the field of risk and safety was established to evaluate the data observed. The group draws conclusions about the safety level, status and trends. In addition, a group of representatives from the various interested parties discuss and review important safety issues, supporting documentation and views of the status and trends in general, as well as the conclusions and findings of the expert group. The details of the project are outside the scope of this book; what is important is the fact that consensus was reached concerning the way to describe risk. In this sense an 'objective' characterization of risk was achieved.

Example. Interval analysis

A system comprises two units in parallel, so that the system functions if at least one of the units functions. Let θ_0 be the unreliability of a unit, that is, the frequentist probability (chance) that the unit is not functioning. Historical data for the units make it reasonable to conclude that $0.01 \leq \theta_0 \leq 0.10$. As a model of the unreliability θ of the system, we use the model $\theta = \theta_0^2$, which is based on the assumption of independence. Based on this model and the assumption that $0.01 \leq \theta_0 \leq 0.10$, we obtain an interval for the system unreliability: $0.0001 \leq \theta \leq 0.01$. If the independence assumption is dropped, we are led to bounds $0 \leq \theta \leq 0.1$, as the unreliability of the system is less than or equal to the unreliability of a specific unit. It is not possible to establish better bounds without making some additional assumptions. □

The analysis can be viewed as objective, but it can be criticized for being somewhat non-informative. The decision-maker may ask for a more refined assessment. The price to pay is loss of objectivity. We refer to the discussion in Aven (2009).

Risk assessment is often used to 'prove' that the risk related to an activity is acceptable. Conclusions are reached based on an 'objective' risk description. However, this type of proof and argumentation has been strongly criticized by many researchers and analysts (see Chapter 18). Aven (2008) concludes that the arbitrariness in the numbers produced could be significant, due to the uncertainties in the estimates or as a result of the uncertainty assessments being strongly dependent on the assessors. 'As experienced risk analysts, we have all struggled with the lack of precision in risk analysis, and it has also been documented by several benchmarking exercises, see e.g. Lauridsen *et al.* (2001)' (Aven, 2008).

To lend substance to this discussion about objectivity, we need to define more precisely the risk and probability perspective adopted. But before we do this, we introduce an example of a risk assessment of a planned process plant.

Example. Risk assessment of a planned process plant

A process plant is planned close to an inhabited area, and a risk assessment is carried out to demonstrate that the accident risk is acceptable. To this end, the operator of the plant has defined a set of risk acceptance criteria (tolerability limits). These include an F–N curve as in Figure 2.1, as well as an individual risk criterion for third party risk, namely that the individual probability of being killed for a person living, working or staying outside the plant shall not exceed 1×10^{-5} for a period of 1 year.

The risk assessment is carried out. Suppose that it produces results showing that these criteria are met. Then the conclusion of the operator is that the risk is acceptable. □

This issue is studied in more detail in Chapter 18. The point we wish to make here is that the risk results are reported as objective. There is no discussion of the dependence of the results on the assumptions made and the analysts used for the risk assessment. It is an example of a study which serves the need of the operator to obtain acceptance of the plant. Its rationale can be questioned as the picture of risk does not exist independently of the assessor and there is no consensus about the picture of risk among all stakeholders. In the process plant case several independent analysts argue that the risk numbers should have been higher than shown in the study report as a key assumption is not valid. This view is supported by the neighbours of the plant. To be more precise on this, we need to define the risk and probability perspective adopted.

Let us momentarily return to our die example. The die has a probability p of giving a 6 and $1 - p$ of giving a 1, 2, 3, 4 or 5. Here p is understood as the fraction of 6s if the die is thrown over and over again. We may interpret p as a property of the die. It exists independent of the analysts, and is thus 'objective'. However, p is unknown and has to be estimated and the estimate is, of course, not objective. You may estimate p by $p^* = 0.90$, but others could estimate p differently. In the absence of strong data supporting a specific estimate, we would expect large differences in individual estimates. Subjective assessments of uncertainties in p may be carried out, but these would, of course, not in general be objective.

For the process plant example the set-up is similar. Objective frequentist probabilities and expected values are defined, for example the individual risk and the F–N curve. In the risk assessment study these parameters are estimated. The estimates are, however, not objective as stressed above.

For a knowledge-based (subjective) probability, it is obvious that the results of the risk assessments are not in general objective. However, according to criterion (ii) we may still refer to the results as 'objective' if consensus can be obtained among all stakeholders. In the process plant example, such consensus is not achieved.

Hence, an objective risk picture cannot in general be obtained. Experts may agree upon the characterization of risk, but other stakeholders may disagree. If there is consensus among all stakeholders about the risk characterization, this does not mean, of course, that they come to the same conclusion when it comes to risk acceptance. One stakeholder could find a level acceptable whereas another could find it intolerable (see Chapter 18).

Probability is understood as knowledge-based (subjective) probability

The underlying objective probability is easy to understand when considering repeated experiments as for the die example. In practice, the meaning of this objective probability is not so clear. Consider the probability of at least one fatality in a year in our process plant. According to the relative frequency view, this probability is interpreted as the proportion of plants with at least one fatality when considering an infinite number of similar plants. This is, of course, a thought experiment – in real life we just have one such plant. Therefore, the probability is not a property of the unit itself, but the population it belongs to. How should we then understand the meaning of 'similar plants'? Does it mean the same type of buildings and equipment, the same operational procedures, the same type of personnel positions, the same type of training programmes, the same organizational philosophy, the same influence of exogenous factors, and so on? As long as we speak about similarities at a macro level, the answer is yes. But something must be different, because otherwise we would get exactly the same output result for each plant, either the occurrence of at least one fatality or no such occurrence. There must be some variation at a micro level to produce the variation of the output result. So we should allow for variations in the equipment quality, human behaviour, and so on. But the question is to what extent we should allow for such variation. For example, in human behaviour, do we specify the safety culture or the standard of the private lives of the personnel, or are these factors to be regarded as factors creating the variations from one facility to another, that is, the stochastic (aleatory) uncertainty? We see that we will have a hard time specifying what should be the framework conditions of the experiment and what should be stochastic uncertainty. In practice, we seldom see such a specification carried out, because the framework conditions of the experiment are tacitly understood. As seen from this example, it is not obvious how to make a proper definition of the population.

Reflection

A recent study performed shows that people underassess their own risk compared to objective risk (Andersson and Lundborg, 2007). The case relates to road-traffic risk. The objective risk is the mortality numbers given for the relevant gender and age group, whereas the risk judgement is the individual's assignment of his/her death probability:

> In an average year the risk of dying in a traffic accident for an individual in her/his 50s is 5 in 100 000. What do you think your own annual risk of dying in a traffic accident will be? Your risk may be higher or lower than the average. Consider how often you are exposed to traffic, what distances you travel, your choice of transportation mode and how safely you drive.
>
> I think the risk is ... in 100 000.

The objective risk is based on statistics from the period 1995–1999. Comment on this study.

Results like this are well known from the risk perception literature. We know that individuals, when assigning probabilities, are influenced by their own experience, their perception of the risk, and we know that they have problems in judging small probabilities (see Chapters 8 and 12). This is, however, not our focus here. We raise a more fundamental problem. Are we making a proper comparison? The objective risk is not risk itself, but some historical data, more or less relevant to the individual. Let us make this somewhat more formal, and assume a traditional statistical framework where underlying objective probabilities exist in a relative frequency sense. Let p_i be the probability that person i is killed during the next year in traffic. Then the expected number of fatalities, μ, is given by

$$\mu = \sum_i p_i,$$

where the sum is over the total population considered. To estimate μ we may use the observed number of fatalities N_1, in the period 1995–1999, or alternatively estimate each p_i. Denoting the estimate of p_i by p_i^*, we obtain the latter estimate of μ, μ^*, by

$$\mu^* = \sum_i p_i^*.$$

In the study, μ^* is compared to N_1, using estimates p_i^* determined by each relevant person. In this framework we may define a bias, as a true reference level, μ, is assumed. One may argue that N_1 is an accurate estimate of μ, but what about using p_i^* to estimate p_i? Do people understand what p_i means? Do you understand what it means? To give an explanation we have to consider an infinite population of similar individuals (or similar years) and interpret p_i as the proportion of this population where the person is killed. That is a hard concept to understand. Many would say it is a rather meaningless quantity. It helps, of course, to use a relative scale as is done in this study. You may, for example, say that the probability is 10 times higher than the probability of an average person (this is your estimate). Or alternatively, using a knowledge-based probability interpretation, which perhaps is a more natural interpretation in this study, you simply say that your subjective probability of being killed is 10 times higher than the observed frequency (5 out of 100 000 in the example). The term *objective risk* should be used with care as there is no objective risk relevant for this person. There is a type of 'objective risk' for the total population as we would be able to predict the total number of fatalities with quite a high level of precision. But there is no objective risk for the individual. The individual subjective probabilities have a future perspective and a number of factors would obviously affect your probability, for example the fact that you will buy a new car next week. It should be no surprise that the $\sum_i p_i^*$ produced would typically be lower than N_1, and that there are gender and age differences. See Andersson and Lundborg (2007).

Summary

Risk assessments produce risk estimates of an assumed 'objective' risk or knowledge-based (subjective) risk assignments, depending on the perspective to risk adopted. In none of the cases can the results of the risk assessments be characterized as objective. Only in some special cases the results exist independently of the assessor or consensus among all stakeholders can be obtained.

References

Andersson, H. and Lundborg, P. (2007) Perception of own death risk. An analysis of road-traffic and overall mortality risks. *Journal of Risk Uncertainty*, **34**, 67–84.

Aven, T. (2008) A semi-quantitative approach to risk analysis, as an alternative to QRAs. *Reliability Engineering and System Safety*, **93**, 768–775.

Aven, T. (2009) On the need for restricting the probabilistic analysis in risk assessments to variability. Accepted for publication in Risk Analysis.

Lauridsen, K., Christou, M., Amendola, A. *et al.* (2001) Assessing the uncertainties in the process of risk analysis of chemical establishments, in E. Zio, M. Demichela and N. Piccinini (eds), *Safety and Reliability. Towards a Safer World*. Proceedings. Vol. **1**. ESREL 2001, Turin, 16–20 September, pp. 592–606.

Vinnem, J.E., Aven, T., Husebø, T. *et al.* (2006) Major hazard risk indicators for monitoring of trends in the Norwegian offshore petroleum sector. *Reliability Engineering and System Safety*, **91**, 778–791.

Further reading

Lauridsen, K., Christou, M., Amendola, A. *et al.* (2001) Assessing the uncertainties in the process of risk analysis of chemical establishments, in *Safety and Reliability. Towards a Safer World*, Proceedings of ESREL 2001, Vol. 1 (eds E. Zio, M. Demichela and N. Piccinini) Politecnico di Torino, Torino, 16–20 September 2001, pp. 592–606.

Aven, T. (2009) On the need for restricting the probabilistic analysis in risk assessments to variability. Accepted for publication in *Risk Analysis*.

12

There are large inherent uncertainties in risk analyses

Consider the following statement from a team of experienced risk analysts about uncertainty in quantitative risk assessment (Aven, 2008):

> The analysis is based on the 'best estimates' obtained by using the company's standards for models and data. It is acknowledged that there are uncertainties associated with all elements in the analysis, from the hazard identification to the models and probability calculations. It is concluded that the precision of the analysis is limited, and that one must take this into considerations when comparing the results with the risk acceptance criteria and tolerability limits.

Based on such a statement one gets the impression that there are large inherent uncertainties in risk analysis. Everything is uncertain, but is risk analysis not performed to assess the uncertainties? From this statement it looks like the risk analysis generates uncertainty.

To discuss the assertion that there are large inherent uncertainties in risk analysis, we need to clarify what we are uncertain about. Is it aspects of the world expressed by quantities (X) such as costs, the occurrence of accidents and the number of fatalities? Or are the uncertainties related to our uncertainties about the world and the risk description (see Figure 12.1)? In the latter case, probabilities P_X are used to describe the risk related to X.

To address this, we consider two possible objectives of the risk analysis: the purpose of the risk analysis is to

1. accurately estimate the risk, that is, the probabilities P_X, or
2. describe the uncertainties about the world (see Figure 12.2).

Misconceptions of Risk T. Aven

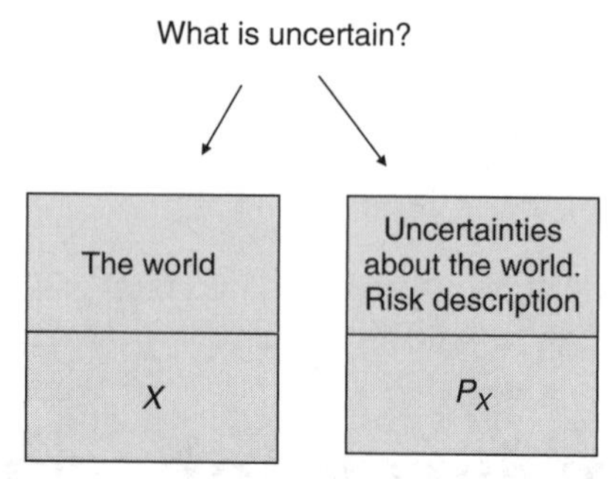

Figure 12.1 Illustration of the issue: what is uncertain in risk analysis? Here X may represent costs or the number of fatalities and P_X the probability distribution of X.

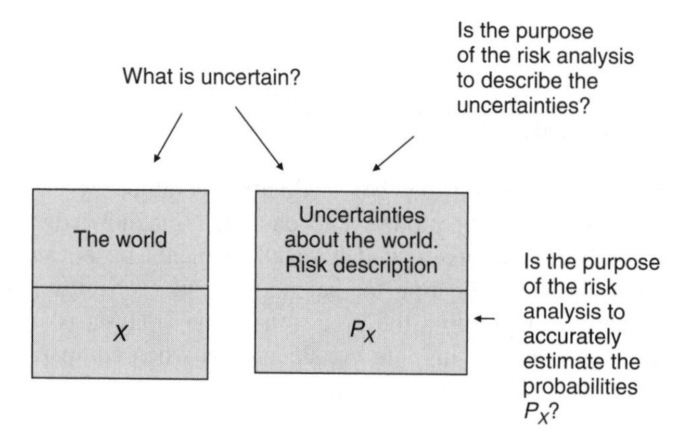

Figure 12.2 Different aims of the risk analysis.

The objective of the risk analysis is to accurately estimate the risk (probabilities)

Suppose that a risk analysis investigates the probability p of an accidental event A. The probability p is defined as the fraction of times the accidental event A occurs if the activity considered were repeated (hypothetically) an infinite number of times. It expresses variations within this mentally constructed infinite population. Since the true value of p is unknown, it must be estimated. The estimation accuracy (precision) can be described in different ways, but the common tool is confidence intervals. We may estimate p directly using a set of relevant data on occurrences of the accidental event A, or indirectly using models. In this latter case we may write $p = G(q)$, where q is a vector of parameters, representing probabilities and expected values on a more detailed system level, and G is the model. The model G may, for example, be based on a set of event trees and fault trees (Bedford and Cooke, 2001). If relevant data exist for q, we can construct estimates q^* of q, and through the model G we obtain an estimate of p, p^*, from the equation $p^* = G(q^*)$.

If we have available a substantial amount of data and these are considered relevant for the estimation of the statistical parameter p – that is, the observations are considered similar to those of the population generating p – statistical theory shows that the estimation error becomes negligible (see Chapter 2).

However, in a practical risk analysis setting, data are often scarce. If the statistical analysis is based on few data, the estimation accuracy would be poor and the confidence intervals would be wide. Hence, if the goal of the risk analysis is to obtain accurate estimates of some true risk, we can quickly conclude that risk analysis fails as a scientific method.

It is not possible to define precisely what we mean by terms such as 'accurate estimation' and 'negligible estimation error' without being explicit about the context. We may, however, indicate an order of magnitude. For example, if we estimate a p equal to 0.10 and the upper and lower confidence bounds are 0.11 and 0.09 respectively, the estimation error would be considered negligible for most applications.

To increase the amount of data, we often extend the relevant population of observations to cover situations that, to a varying degree, are similar to the one being studied. This reduces the quality, that is, the relevance of the data, but this aspect would not be possible to describe by the statistical analysis. If the data are not considered relevant, the statistical analysis cannot be used to check that the risk numbers produced are accurate compared to the underlying true risk.

The same type of problems arise in the case of modelling, although the amount of data is often larger at the detailed system level. However, in this case we should also take into account the uncertainty (inaccuracy) introduced by using the model G. In the analysis we use G, and this introduces a possible error in the risk estimation as $G(q)$ could deviate from p. Hence, accurate estimation is ensured only if the model uncertainty (inaccuracy) is negligible (see Chapter 13).

Example

Consider a situation characterized by large uncertainties. Think of the probability p that a complex system which is based on new technology will function as required in a specific year. It is a unique system and few relevant data exist beyond test observations for some critical subsystems. Or think of the probability q that the value of a specific security will increase by a minimum of 5% next year. Data are plentiful in this case, but the question is to what extent these data are relevant. Based on the historical observations we may estimate a p^* equal to 0.75 (say), but the estimate is subject to large uncertainties. Estimates of the underlying probabilities can always be determined, but it is difficult to know how accurate these estimates are.

In both cases these systems (activities) are rather unique and there are large uncertainties about the future performance of the system (security). However, an infinitely large population of similar systems (activities) needs to be introduced to define the frequentist probabilities p and q. In these populations p and q represent the fraction of systems (activities) where the event of interest occurs. The concept is abstract and its meaning is not obvious, as discussed in Chapter 2. The statistical analysis is, however, not concerned with the interpretation of the probabilities, but with the estimation of these quantities. With a lot of relevant data we can accurately estimate the underlying true probabilities. In our case, however, this is not possible as such data are not available.

Suppose that we have four observations of similar systems and one indicates failure. The natural estimate of p is then $p^* = 1/4$. This estimate is obviously subject to large uncertainties as the number of observations is so few. The true p could be 'any number'. We could try to establish a confidence interval for p, but it would lead to a meaningless wide interval [0,1]. If we observed no failures, an estimate $p^* = 0$ would be derived. This estimate is obviously too low, but the data provide no arguments for an alternative estimate. The use of modelling could improve the estimation, as discussed above. At the lower system level we may have more data available, and if the model G is a good description of the world, we have obtained an improved method for estimating p. The problem is, of course, that in the case of large uncertainties about the performance of the system, it may be difficult to establish such a model G. How can we know that the model is an accurate description of the real world? Testing at the equipment level is possible, but at an overall system level may be impossible. Hence, we may establish accurate estimates of p provided that G is a good model. If G is poor, this alternative approach has not led to an improved estimation of p, perhaps the opposite. However, the modelling may have given useful insights into the performance of the system. □

Fortunately, in many cases it is possible to define a large population of similar systems (activities). Think, for example, about a population of human beings in a country. Let r be the proportion of this population having a certain property (e.g. symptoms of a type of illness). Now suppose that this fraction r is unknown, that there is a lack of knowledge about the frequency of this property in the population. Can we then express the probability that an arbitrarily selected person from this population has this property? No, this probability equals r and is unknown, but we can estimate it accurately by taking a sample of human beings and checking whether they have this property or not. This would be a standard application of statistical analysis and we may use confidence intervals to express the uncertainties (see Chapter 2). However, without data the estimates would be subject to large uncertainties, and the associated risk would not have been determined.

Reflection

Suppose you are a member of this population of human beings. Are r and the associated estimate r^* (based on a sample from this population) adequate measures of risk for you?

They are adequate measures of risk to the extent that the average performance in this population is relevant for you. If you are a typical member of the population the average may be a very informative measure. However, if you are not a typical member of the population, the average could provide little or no information relevant for you. Defining subgroups may lead to more relevant groups, but there is a balance to be struck, as more specific groups reduce the amount of data available.

The objective of the risk analysis is to describe our uncertainties about the world

If uncertainty description is the objective of the risk analysis, the statement that there are large uncertainties in risk analysis is not valid. There are not large uncertainties in risk analysis, but there could be large uncertainties about the world, about the occurrence of events and their consequences, as well as the underlying factors causing these events and consequences to occur. The aim of the analysis is to describe these uncertainties. This means that we have to

(a) identify what is uncertain, and

(b) specify who is uncertain.

Often we are able to reduce (a) to a set of quantities, such as costs, the occurrence of events, and the number of fatalities. We denote these quantities by X. In most cases in a risk analysis the answer to (b) is the analysts. However, in some situations it could be essential to present different uncertainty assessments reflecting the different views among experts and various stakeholders. Integrating all views into one aggregated figure could camouflage important aspects of the risk picture.

Next, we have to address the issue of how to express the uncertainties. Probability is the common tool, but a probability has different interpretations. We will, however, argue, in line with the discussion in Chapter 2, that to describe the uncertainty, the probabilities must be knowledge-based (subjective) probabilities with reference to a standard expressing the analysts' uncertainty about unknown quantities X. Following this interpretation, the assessor compares his/her uncertainty about the occurrence of the event A with the standard event of drawing at random a favourable ball from an urn that contains $\mathrm{P}(A)$ favourable balls.

Frequentist probabilities P_{f} are not measures of uncertainty. They exist as proportions of infinite or very large populations of similar units to those considered. Such proportions are referred to as chances in the Bayesian terminology (Singpurwalla, 2006). They are treated as unknown quantities X.

The uncertainty descriptions P cannot be frequentist probabilities P_{f} as such probabilities are, in fact, not somebody's measure of uncertainty, but a way of expressing variation within a real or thought-constructed infinite (or very large) population of similar units to those studied. As an example, let P_{f} be the frequentist probability that a technical component fails during a specific period of time. This 'probability', which is understood as the fraction of components that fail in this period when considering an infinite large population of similar components, is unknown. Hence, P_f describes an unknown population fraction. The result is that P_f has to be considered in line with X, as unknown quantities subject to uncertainties. Consequently, P needs to be interpreted as knowledge-based (subjective) probabilities.

These probabilities express *epistemic uncertainties*. The variation in the populations of similar units to the one studied, that for example generates the true value of P_f, is often referred to as *aleatory (stochastic) uncertainty*. This uncertainty is, however, not an uncertainty for the analysts.

But a knowledge-based (subjective) probability can be given different interpretations, as was noted in Chapter 2: the betting interpretation and the reference to a standard. Following the betting interpretation, the probability of the event A, P(A), equals the amount of money that the assigner would be willing to bet if he/she received a single unit of payment if the event A were to occur, and nothing otherwise (Singpurwalla, 2006). And as concluded in Chapter 2, to describe uncertainties, this interpretation is not appropriate, as it extends beyond the realm of uncertainty assessments – it reflects the assessor's attitude to money and the gambling situation, which means that analysis (evidence) is mixed with values. The scientific basis for risk analysis is based on the idea that professional analysts describe risk in isolation from how we (the assessor, the decision-maker or other stakeholders) value the consequences and the risk.

Different probability-based measures are used to describe risk, such as expected values, variance and quantiles. But a full risk description needs to see beyond these measures of P. All probabilities are conditional on a background knowledge K, which includes assumptions and suppositions, and in particular the models used in the analysis. This background knowledge is an integral part of the results of the analysis and all probabilities need to be considered in relation to K. Several examples were considered in Chapter 2. Here is another one related to costs. To support a decision on investing in a project, a risk assessment is performed. The assessment is based on the assumption that the oil price is \$100 per barrel. The probabilities produced are conditional on this assumption. However, the oil price is an uncertain quantity. The actual oil price could deviate strongly from \$100. This uncertainty could be included in the assessment, by specifying a knowledge-based probability distribution for the oil price. But this distribution would again be conditional on a set of assumptions, for example related to the database used to determine the distribution or some underlying assumptions about how the future will be, compared to the historical observations. The analysts need to clarify what is uncertain and subject to the uncertainty assessment and what constitutes the background knowledge. From a theoretical point of view one may think that it is possible (and desirable) to remove all such uncertainties from the background knowledge, but in a practical risk assessment context that is impossible. We will always base our probabilities on some type of background knowledge, and often this knowledge would not be as easy to specify as the oil price.

The uncertainty assessments of these factors could, for example, take the following form (Aven, 2009): each factor's importance is measured using a sensitivity analysis. Is changing the factor important for the risk indices considered, for example the probability that the loss exceeds a specific number? If this is the case, we next address the uncertainty of this factor. Are there large uncertainties about this factor? If the uncertainties are assessed as large, the factor is given a high risk score. Hence, to obtain a high score in this system, the factor must be judged as important for the risk indices considered and the factor must be subject to large uncertainties. This uncertainty assessment goes beyond the probabilistic analysis.

As an example, think of an assessment which is based on the following assumptions:

- The use of a gas leakage database L which includes data from plants operated under quite different operating conditions from the one considered
- The use of a new technology for operating a system (S) is not changing the overall performance of the system compared to average industry data.

From this list we identify two uncertainty factors:

(i) the number of gas leakages

(ii) the performance of the system S.

An assessment is performed of these factors, according to the ideas outlined above. The risk indices are sensitive to changes in these factors, and both factors are subject to considerable uncertainties. Special concern is related to the use of the database L, and based on the risk evaluation it was decided to look more closely into this factor and to try to reduce the uncertainties.

What we suggest is a broad risk assessment of the uncertainty factors, not a detailed quantitative risk assessment but a more qualitative approach. Crude probabilities may be assigned for the uncertainty factors but these do not replace the qualitative judgements. The search for quantitative, explicit approaches for expressing the uncertainties, even beyond the knowledge-based (subjective) probabilities, may seem to be a possible way forward. However, such an approach is rejected. Trying to be precise and accurately expressing what is extremely uncertain does not make sense. Instead, we recommend a more open qualitative approach for revealing such uncertainties, as described above.

Knowledge-based (subjective) probabilities

To specify a knowledge-based probability, different approaches can be used (Aven, 2003):

Derivation of an assigned distribution based on classical statistics. For example, if we have 3 'successes' out of 10 observations, we obtain $P(A) = 0.3$, where A is the 'success' event. This is our (i.e. the analyst's) assessment of uncertainty about A. This probability is not an estimate of an underlying true probability $p = \mathrm{P}(A)$ as in the classical setting, but an assessment of uncertainty related to the occurrence of A. Thus $\mathrm{P}(A) = 0.3$, whereas in the classical setting, $P^* = 0.3$, where P^* is an estimate of p, that is, $p \approx 0.3$, hopefully.

This method is appropriate when the analyst judges the observational data to be relevant for the uncertainty assessment of A, and the number of observations, n, is large. In cases with large uncertainties, this method would not be feasible.

Analyst judgement using all sources of information. This is a method commonly adopted when data are absent or are only partially relevant to the assessment endpoint. A number of uncertain exposure and risk assessment situations are in this category. The responsibility for summarizing the state of knowledge, producing the written rationale, and specifying the probability distribution rests with the analyst.

The starting point for the specification of $\mathrm{P}(A)$ is that the analyst is experienced in assigning probabilities expressing uncertainty, so that he/she has a number of reference points – he/she has a feeling for what 0.5 means in contrast to 0.1, for example. To facilitate the specification he/she may also think of some type of replication of similar events as generating A, and think of the probability as corresponding to the proportion of 'successes' that he/she would predict among these events. For example, suppose that he/she predicts 1 'success' out of 10; then he/she would assign a probability 0.1 to $\mathrm{P}(A)$. Note that this type of reasoning does not mean that the analyst presumes the existence of a true probability, it is just a tool for simplifying the specification of the probability.

To specify a probability distribution of X, which is a random quantity on the real line, the common approach is to assign probabilities as above for the events $X \leq x$, or $X > x$, for suitable numbers x. Often one starts with a percentage, say 90%, and then specifies the value x such that $\mathrm{P}(X > x) = 0.90$. Combining such quantile assessments with a specified distribution class, such as the normal distribution or a lognormal distribution, only a few assessments are needed (typically two, corresponding to the number of parameters of the distribution class).

An alternative approach for the specification of $\mathrm{P}(X \leq x)$ is to use the maximum entropy principle. This approach means specification of some features of the distribution, for example the mean and variance, but not the whole distribution. Then a mathematical procedure gives a distribution with these features and, in a certain sense, minimum information beyond that. We refer to Bedford and Cooke (2001), p. 73.

To specify the probability distribution, the analyst may consult experts in the subject of interest, but the uncertainty assessment is not a formal expert elicitation as explained below.

Formal expert elicitation. This approach requires the analyst to identify and bring together individuals acknowledged as 'experts' in the subject of concern. A typical procedure is the following. The analyst trains the experts in the assessment problem and disseminates among the experts all relevant information and data. The experts are then required to formalize and document their rationales. They are interviewed and asked to defend their rationales before committing to any specific probability distribution. The experts specify their own distribution by determining quantiles.

In some cases, weights are assigned to the experts to distinguish differences in 'expertise'. Some argue that that the selection of high-quality experts at the outset is mandatory and that all experts used for the final elicitation should be given the same weight. Others argue that the experts be given the opportunity to assign weights to themselves.

The method is extremely difficult to rebuke, except by conducting new experiments on the uncertain quantity of interest or convening a separate independent panel of experts.

It is a basic principle of the risk analysis that the analyst is ultimately responsible for the assessment, and as such, the analyst is obliged to make the final call on the probability distribution. 'Experts' have advanced knowledge in rather narrow disciplines and are unlikely to devote the time necessary (even with training) to become as familiar with the unique demands of the assessment question as is the analyst. However, the experts' distributions should not be changed if this possibility is not a part of an agreed procedure for the elicitation between the analyst and the experts.

Formal expert elicitation may be undertaken when little relevant data can be made available and when it is likely that the judgement of the analyst will be subject to scrutiny, resulting for example in costly project delays. Formal expert elicitation could be very expensive, so a justification for when to adopt such a procedure is required.

Experts may specify their own probability distributions, or they could provide the analyst with information for him/her to process and finally transform to a probability distribution. This latter approach has the advantage that the expert can speak his own language and avoid the somewhat abstract formalism of using probabilities. On the other hand, it may be difficult for the analyst to fully understand the expert judgements if they are just reports of knowledge, with no reference to the probability scale.

Building consensus, or rational consensus, is of major concern when using expert opinions. Five principles are often highlighted (Cooke, 1991):

- **Reproducibility:** It must be possible to reproduce all calculations.
- **Accountability:** The basis for the probabilities assigned must be identified.
- **Empirical control:** The probability assignments must in principle be susceptible to empirical control.
- **Neutrality:** The methods for combining or evaluating expert opinion should encourage experts to state their true opinions.

- **Fairness:** All experts are treated equally, prior to processing the results of observations.

A remark on 'empirical control' is in order. Empirical control does not apply to the probability at the time of the assignment. When conducting a risk analysis we cannot 'verify' an assigned probability, as it expresses the analyst's uncertainty prior to the observation.

The Bayesian approach gives a unified approach to the specification of P(A) and P($X \leq x$). The above three approaches may also be viewed as Bayesian, although they are to a large extent based on direct probability assignments without introducing a probabilistic model as is in the common Bayesian framework (see Chapter 15).

Challenges related to the specification of knowledge-based probabilities

In the case of little relevant data it may be difficult to determine the probability. You specify a probability equal to 0.90 but you may feel that it is hard to distinguish between 0.90 and 0.80 (say). Following the perspective of knowledge-based probabilities there is no reference to a correct probability, but would this mean that we cannot speak about the 'goodness' of the probability assignments?

A knowledge-based probability is a subjective measure of uncertainty related to an event or, more generally, an observable quantity X, based on the assessor's state of knowledge. In principle an observable quantity can be measured, thus probability assignments can, to some extent, be compared to observations. We write 'in principle' as there could be practical difficulties in performing such measurements (we are often performing an analysis of thought-constructed systems). Of course, one observation as a basis for comparison with the assigned probability is not very informative in general, but in some cases it is also possible to incorporate other relevant observations and thus give a stronger basis. As noted above, 'empirical control' does not, however, apply to the probability at the time of the assignment. When conducting a risk analysis we cannot 'verify' an assigned probability, as it expresses the analyst's uncertainty prior to observation. What can be done is a review of the background knowledge used as the rationale for the assignment, but in most cases it would not be possible to explicitly document all the transformation steps from this background knowledge to the assigned probability.

Thus, a traditional scientific methodology based on empirical control cannot be applied for evaluating such probabilities. It is impossible in general to obtain repeated independent measurements of assigned probabilities from the same individual because he/she is likely to remember his/her previous thoughts and responses. Consequently, there are no procedures for the measurement of the probability assignments that permit the application of the law of large numbers to reduce 'measurement errors'.

The difficulties involved in applying standard measurement criteria of reliability and validity to the measurement of probability assignment give rise to the question of how to evaluate and improve such assignments. Three types of criteria, referred to as syntactic, pragmatic and semantic (calibration), have been suggested (Lindley *et al.*, 1979; see also Aven, 2003).

Syntactic criterion

The syntactic criterion is related to the probabilities obeying syntactic rules – the relations between assignments should be governed by the laws of probability. For example, if A and B are disjoint events, then the assigned probability of the event A or B should be equal to the sum of the assigned probabilities for A and B. A set of probability assignments is (internally) coherent only if it is compatible with the probability axioms. Coherence is clearly essential if we are to treat and manipulate the probability assignments.

Pragmatic criterion

The pragmatic criterion is based on comparison with 'objective' values, the reality, and is applicable whenever the assigned probability of an event, for example a royal flush in poker, or a disease, can be meaningfully compared to a value that is computed in accordance with the probability calculus or derived from empirical data. For example, if history shows that out of a population of a million people, about two suffer from a certain disease, we can compare our probability to the rate $2/10^6$. However, such tests cannot be applied in many cases of interest as such proportions cannot be specified and sufficient relevant data are not available. Thus, the pragmatic criterion is only relevant in special cases.

Calibration

Calibration tests relate to the ability to obtain correct statements when considering a number of assignments. Formally, a person is said to be well calibrated if the proportion of correct statements, among those that were assigned the same probability, matches the stated probability, that is, his hit rate matches his confidence. Clearly there is no way of 'validating', for example, a risk analyst's single judgement that the probability of a system failing during a one-year period of operation is 0.1. If, however, the analyst is assessing many systems, we would expect system failure to occur about 10% of the time. If, say, 50% of the systems fail, he is badly calibrated. Often a scoring rule is used to reward a probability assessor on the basis of later observed outcomes. A simple scoring rule is the quadratic rule. If you assign a probability p for an event A, this rule gives the score $(1 - p)^2$ if the event is true and p^2 if it is false.

Again, it is difficult to apply the criterion. The problem is that it does not apply at the point of assessment. The probability assignments are supposed to provide decision support, so evaluation of the 'goodness' of the probabilities needs to be made before the observations, as already noted. The probabilities are assigned for alternative contemplated cases, meaning that comparisons with observations would be possible for just some of the probabilities assigned. There could also be changes in the background conditions of the probabilities from the assignment point to the observations. In risk analysis applications it often takes a long time before observations are available. The probabilities are, in many cases, small (rare events), which means that it is difficult to establish meaningful hit rates. Suppose that we categorize probabilities in groups of magnitude 0.1 and 0.01 only. And suppose that we observe that for the two categories the risk analyst obtains one 'success' out of 20 cases, and zero out of 50 cases, respectively. Is the risk analyst then calibrated? Or to what extent is he/she calibrated? The hit rate for the first situation is 0.05, just a factor of 2 below his/her confidence, whereas in the second situation, the hit-rate is zero, which makes it difficult to compare with the probability assignments.

We conclude that calibration in general is not very useful for evaluating the 'goodness' of the probability assignments in a risk analysis. Rather we see calibration as a tool for training risk analysts and experts providing input to the risk analysis, in probability assignments. By considering situations of relevance to the problems being analysed and where observations are available, we can evaluate the performance of the analysts and the experts, and improve their calibration in general. Such training would increase the credibility of the risk analyst and the experts providing input to the risk analysis.

In situations where a number of probabilities are assigned and observational feedback is quick, such as in weather forecasting, comparisons with the observed values provide a basis for evaluating the 'goodness' of the assessors and the probabilities assigned. In addition to calibration, several other characteristics of prediction performance are useful, such as refinement or sharpness. Refinement relates to a sample of probability assignments and is defined as the degree to which the assignments are near zero or one (Murphy and Winkler, 1992). A well-calibrated assessor need not be a good predictor or forecaster. If the relative rate of an event A is 30%, the assessor would be well calibrated if he assigns a probability of A of 30% all the time. The refinement would, however, be poor.

Heuristics and biases

People tend to use rather primitive cognitive techniques when assigning probabilities, that is, so-called heuristics. Heuristics for assigning probabilities are easy and intuitive ways to deal with uncertain situations. The result of using such heuristics is often that the assessor unconsciously tends to put 'too much' weight on insignificant factors. Some of the most common heuristics are the following (Tversky and Kahneman, 1974):

- The assessor tends to base his probability assignment on the ease with which similar events can be retrieved from memory. Occurrences of events where the assessor can easily retrieve similar events from memory are likely to be given higher probabilities than occurrences of events that are less vivid and/or completely unknown to the expert (availability heuristic). A classic example is a person who argues that cigarette smoking is not unhealthy because his grandfather smoked three packs of cigarettes a day and lived to be 95.

- The assessor tends to choose an initial anchor. Then extreme points are assessed by adjusting away from the anchor. One of the consequences is often a low probability of extreme outcomes (anchoring and adjusting heuristics). A well-known example of this heuristic is provided by Tversky and Kahneman (1974) in one of their first studies. They showed that when asked to estimate the percentage of African nations which are members of the United Nations, people who were first asked if it was more or less than 45% gave lower estimates than those who had been asked if it was more or less than 65%. The anchor here is the reference percentage in the first question.
- The assessor assigns a probability by comparing his knowledge about the phenomenon with the stereotypical member of a specific category. The closer the similarity between the two, the higher the judged probability of membership in the category (representativeness heuristic). This heuristic can be illustrated by the following example (Tversky and Kahneman, 1974). Consider an individual who has been described by a former neighbour as follows: 'Steve is very shy and withdrawn, invariably helpful, but with little interest in people, or in the world of reality. A meek and tidy soul, he has a need for order and structure, and a passion for detail.' How do people assess the probability that Steve is engaged in a particular occupation from a list of possibilities (e.g. farmer, salesman, airline pilot, librarian or physician)? How do people order these occupations from most to least likely? In the representativeness heuristic, the probability that Steve is a librarian, for example, is assessed by the degree to which he is representative of or similar to the stereotype of a librarian.

Daniel Kahneman and Amos Tversky published a series of seminal articles in the field of judgement and decision-making, including their *Science* article of 1974. Kahneman was awarded the Nobel Prize in Economics in 2002 and in his autobiography (Kahneman, 2002) he writes:

> The *Science* article turned out to be a rarity: an empirical psychological article that (some) philosophers and (a few) economists could and did take seriously. What was it that made readers of the article more willing to listen than the philosopher at the party? I attribute the unusual attention at least as much to the medium as to the message. Amos and I had continued to practice the

> psychology of single questions, and the *Science* article – like others we wrote – incorporated questions that were cited verbatim in the text. These questions, I believe, personally engaged the readers and convinced them that we were concerned not with the stupidity of Joe Public but with a much more interesting issue: the susceptibility to erroneous intuitions of intelligent, sophisticated, and perceptive individuals such as themselves. Whatever the reason, the article soon became a standard reference as an attack on the rational-agent model, and it spawned a large literature in cognitive science, philosophy, and psychology. We had not anticipated that outcome.

The training of the risk analyst and of the experts providing input to the risk analyst should make them aware of these heuristics, as well as other problems of quantifying probabilities such as superficiality and imprecision which relate to the assessors' possible lack of feeling for numerical values. Lack of precision is particularly a problem when evaluating events at the lower part of the probability scale, typically less than 1/100.

Although all experts seem to have a limit probability level under which expressing uncertainty in numbers becomes difficult, practice shows that this limit can be lowered by training. By repeatedly facing the problem of assigning probabilities to rare but observed events (the result, of course, not known to the analyst or expert a priori), discussing the causal factors and comparing their likelihood, the analyst or expert familiarizes himself with this way of thinking. The analyst or expert will gradually feel more comfortable with applying smaller numbers, but still training alone will hardly solve this problem. It seems that we must accept that the area of application of probability judgement has a boundary at the lower and upper ends of the probability scale in which probability assignments have low confidence.

Given this fact, the challenge is to design models such that the number of low (high) probability events to be specified is minimized. Instead of specifying p we introduce a model G and specify probabilities for the components (arguments) of G. From this uncertainty propagation we are led to a probability of the event of interest. We refer to the earlier discussion covering the traditional statistical basis.

Evaluation of the assessors

We consider a risk analyst who wishes to specify a probability distribution $P(X \leq x)$. This probability is a measure of uncertainty; it is not an observable quantity. No true value of $P(X \leq x)$ exists. Consequently, we cannot draw a conclusion on the correctness of this probability distribution as such. Of course, if the pragmatic criterion applies – that is, the probabilities can be compared to 'objective' values – assessors can be meaningfully evaluated. For example, if an analyst predicts two failures of a system during a period of 1 year, and the associated uncertainty is considered negligible, this assessment and the assessor would be judged as poor if there is strong evidence showing that such systems would fail at least ten times a year. Unfortunately, the pragmatic criterion does not often apply in a risk analysis context. Sufficient relevant data do not exist. The 'goodness' of the numerical probability is then more a question about who is expressing their view, what competence they have, what methods and models they use and their information basis in general, as well as what quality assurance procedures have been adopted in the planning and execution of the assessment. Thus, we make a clear distinction between the numerical probability itself, which cannot be validated, and the evaluation of the number and the basis for it. Confidence in the probability assignment process is essential. This confidence is affected by several factors:

- The gap judged by the evaluator of the probability (who could be the decision-maker), between the assessor's state of knowledge and the 'best information available'.
- The evaluator considers the 'best information available' to be insufficient.
- Motivational aspects.
- The training of the assessor in probability assignments, and in particular how to treat heuristics and 'biases', superficiality and imprecision, as discussed above.

If the evaluator considers the assessor's level of information (knowledge) to be significantly lower than the 'best information available', he/she would find the results of the analysis not very informative. The evaluator will be sceptical of the assessor as an expert. Trying to use the best expertise available does not fully solve this problem since in practice there will always be time and cost constraints. Of course, even if the analyst or expert is considered to have the 'best information available', there could be a confidence problem. The evaluator may judge the 'best information available' to be insufficient, and further studies are required to give a better basis for the probability specification.

A risk analyst (expert) may assign a probability that completely or partially reflects inappropriate motives rather than his deeply felt belief regarding a specific event's outcome. As an example, it is hard to believe that a sales representative on commission would make a completely unprejudiced judgement of two safety valves, one of which belongs to a competitor firm. Another example is an engineer who has been involved in the design process and later is asked to judge the probability of failure of an item he personally recommended to be installed. The engineer claims that the item is 'absolutely safe' and assigns a very low failure probability. The management may reject the sales representative's judgement without much consideration since they believe that inappropriate motives have influenced his judgement. The engineer's judgement might not just as easily be rejected since he obviously is a company expert in this area. On the other hand, incentives are present that might affect his probability specification.

Motivational aspects will always be an important part of evaluating probabilities and thus the usefulness of analyses that include expert judgements. In general we should be aware of the existence of incentives that in some cases could significantly affect the assignments.

Reflection

Are motivational aspects a problem when professionals perform risk analysis?

No, in general professional analysts would not intentionally perform a biased assessment, influenced by motivational factors. Their jobs would not last long if their reputation were questioned. However, their approach to the assessment and the methods used could be strongly in favour of one specific party. For example, when performing a standard risk analysis of a process plant, one may argue that important uncertainty factors are camouflaged (see Chapter 18). Do the analysts do anything about this? Do they report on this? Probably not, as it is not in the interest of the client (plant operator). Thus, indirectly, motivational aspects are an important issue when assessing the results of risk assessments.

Standardization and consensus

When conducting many risk analyses within (say) a company there is a need for standardization of some of the probabilities to be used in the analysis, for example related to the distribution of the time to failure of a unit, to reduce the analysis work and ensure consistency. Such standardization requires consensus among the various assessors in the company. However, care has to be taken when searching for such consensus. Different views could be camouflaged and too narrow a risk picture presented.

Reflection

If two persons have the same background information, would that mean that they have the same uncertainties, and therefore the same probabilities?

No, a knowledge-based probability is a judgement made by a specific person. Of course, often we would experience similar numbers if the background knowledge is about the same, but there are no formal constraints in the specification of a knowledge-based probability, implying that my judgement should be the same as yours if we have the same knowledge. A probability is a judgement, and there is no strict mechanical procedure producing one 'correct' value. □

Reflection

The use of knowledge-based probabilities means that there is only one type of uncertainty (epistemic uncertainty). However, it is often felt that some probabilities are easy to assign and feel sure about, others are vague and it is doubtful that the single number means anything. Should not the vagueness be specified?

To provide a basis for addressing this, let us look at an example (Lindley, 1985). A coin is tossed and the event A denotes that it shows heads. In another example, we test a new drug and the event B denotes that it is better than the old drug with a particular pair of patients (what is meant by 'better' is well defined and is not an issue here). In the absence of any information on the type of coin, we would assign a probability to A equal to $1/2$, and this probability is firm in that we would almost all be happy with it. With the drug test we would have an open mind about its effectiveness and similarly ascribe a probability of B equal to $1/2$. The latter value of $1/2$ is vague and one does not feel as sure about it as with the coin. It seems that we have one firm 'objective' probability of $1/2$ and one vague and subjective probability of $1/2$.

Theses two examples put the focus on the background knowledge of the probabilities and the available knowledge to be used as a basis for assessing the uncertainties. Clearly, we know more about the process leading to heads in coin tossing than in the drug example. If we consider 1000 throws, we would be quite sure that the proportion of heads, which we denote p, would be close to $1/2$. Most people would assign very low probabilities for observing (say) less than 100 heads. In the drug example we would, when considering 1000 pairs of patients, have less information about the result, q, representing the proportion of the 1000 pairs of patients benefiting more from the new drug than the old. The new drug could be a complete flop and the old cure is vastly to be preferred, meaning that we would also assign a rather high probability to low values of q. Both low and high values of q are much more probable than low and high values of p, simply because we know that coins cannot easily be that biased, whereas drugs could well be quite different. These different probabilities reflect vagueness and firmness that are respectively associated in our minds with the original probabilities. In the coin example, the background information is so strong that observations would not easily change our assessment, whereas in the drug example, medical evidence would probably lead us to believe in the effectiveness of the new drug. This can be shown formally using Bayes' theorem for updating probabilities (Chapter 15).

This example demonstrates the importance of giving attention to appropriate performance measures. Here they are not A and B, but p and q. When evaluating probabilities in a decision-making context, we always need to address the background knowledge, as it provides a basis for the evaluation.

The need to look beyond the probabilities to express risk

In Chapter 2 we argued that risk is more than probabilities. There are uncertainties to be considered beyond the probabilistic world. Your probability assignment is based on a background knowledge that could hide uncertainties. The assignments may, for example, be based on an assumption that the future performance of a system is similar to the historical observations. This assumption may, however, turn out to be wrong. This type of uncertainty should be reflected when expressing risk. Although this uncertainty is not described by probabilities, it constitutes knowledge-based characterizations. These characterizations are subjective like the probabilities and subject to the same type of problems, for example lack of ability to perform empirical validation. Yet, they express risk.

Reflection

It is common to hear statements saying that risk and probability cannot be expressed in the case of large uncertainties. Argue why this statement is a misconception.

Risk analyses can always be carried out. Risk assignments and risk estimates can always be produced, regardless of access to input data. Through the risk analysis, the knowledge and lack of knowledge one has concerning various quantities may be expressed, and this can always be done. Of course, in a case of large uncertainties, it will be difficult to establish good predictions or estimates, but if the purpose of the analysis is to describe the uncertainties this would not disqualify the analyses. If, on the other hand, the aim of the analysis is to accurately estimate some underlying true probabilities, the probabilities are indeed difficult to determine in the case of large uncertainties. □

Reflection

Are the above challenges for knowledge-based probabilities also relevant to the estimation of frequentist probabilities?

Yes. Consider, for example, the availability heuristic and the person who argues that cigarette smoking is not unhealthy because his grandfather smoked three packs of cigarettes a day and lived to be 95. The heuristic could easily lead to the person providing too low an estimate of the probability of dying due to cigarette smoking.

Summary

It is a common conception that there are large inherent uncertainties in risk analysis. If the aim of the risk analysis is to obtain accurate estimates of some true underlying probabilities and risk, this conception is real in the sense that there are often large uncertainties about the fictional probabilities and expected values produced by the risk analysis. If, however, the aim of risk analysis is to describe uncertainties of observable quantities (X), this conception is a misconception as there are not large uncertainties inherent in the risk analysis, but there are large uncertainties about phenomena and processes in the real world: X is uncertain but not $P(X \leq x)$ and $E[X]$. There are a number of challenges related to the assignment of (knowledge-based) probabilities (as well as estimation of frequentist probabilities), for example related to the use of expert judgements.

References

Aven, T. (2003) *Foundations of Risk Analysis*, John Wiley & Sons, Ltd, Chichester.

Aven, T. (2008) *Risk Analysis: Assessing Uncertainties beyond Expected Values and Probabilities*, John Wiley & Sons, Ltd, Chichester.

Aven, T. (2009) A new scientific framework for quantitative risk assessments. *International Journal of Business Continuity and Risk Management*, **1**, 67–77.

Bedford, T. and Cooke, R. (2001) *Probabilistic Risk Analysis: Foundations and Methods*, Cambridge University Press, Cambridge.

Cooke, R. (1991) *Experts in Uncertainty*, Oxford University Press, New York.

Kahneman, D. (2002) Autobiography. The Sveriges Riksbank Prize in Economic Sciences in Memory of Alfred Nobel 2002, http://nobelprize.org/nobel_prizes/economics/laureates/2002/kahneman-autobio.html.

Lindley, D.V. (1985) *Making Decisions*, John Wiley & Sons, Ltd, London.

Lindley, D.V., Tversky, A. and Brown, R.V. (1979) On the reconciliation of probability assessments. *Journal of the Royal Statistical Society A*, **142**, 146–180.

Murphy, A.H. and Winkler, R.L. (1992) Diagnostic verification of probability forecasts. *International Journal of Forecasting*, **8**, 435–455.

Singpurwalla, N. (2006) *Reliability and Risk. A Bayesian Perspective*, John Wiley & Sons, Ltd, Chichester.

Tversky, A. and Kahneman, D. (1974) Judgement under uncertainty: heuristics and biases. *Science*, **185**, 1124–1131.

Further reading

Aven, T. (2008) Some reflections on uncertainty analysis and management. Accepted for publication in *Reliability Engineering and System Safety*.

Tversky, A. and Kahneman, D. (1974) Judgment under Uncertainty: Heuristics and Biases. *Science*, **185**, 1124–1131.

13

Model uncertainty should be quantified

A model is a representation of the world. It is introduced to obtain insights into the phenomena being studied. The model should describe the world sufficiently accurately, but also simplify complicated features and conditions. There is always a balance to be struck between these concerns. Let us look at a simple example.

Example. Parallel system

We consider a parallel system of two components, as shown in Figure 13.1. The state of the system is denoted Z, and is equal to 1 if the system is functioning and 0 otherwise. The system is functioning if at least one of the components is functioning. The parallel system defines a model G by

$$G(X) = 1 - (1 - X_1)(1 - X_2),$$

where X_i is the state of component i, $i = 1, 2$, defined as 1 if component i is functioning and 0 otherwise.

The true state of the system Z is not identical to $G(X)$; it may turn out, for example that $G(X) = 0$ but $Z = 1$. But in our analysis we choose to use G as a model for Z, and we set Z equal to $G(X)$; that is, $Z := G(X)$.

Assuming independence between the component states, we obtain

$$\mathrm{P}(Z = 1) = \mathrm{E}[Z] = 1 - (1 - p_1)(1 - p_2),$$

where $p_i = \mathrm{P}(X_i = 1)$; that is, p_i equals the probability that component i is functioning. □

Misconceptions of Risk T. Aven

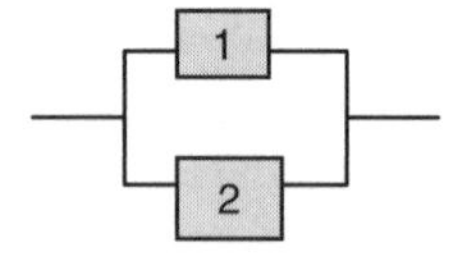

Figure 13.1 A parallel system comprising two components.

We return to the general discussion. The aim is to quantify the analysts' epistemic uncertainties about an unknown quantity Z, which could represent a production volume, profit, a number of fatalities, the fraction of units with a special property in a population, and so on. To this end a model $G(X)$ is introduced, linking some input quantities X to Z as in the example above. Knowledge-based probabilities P are used to express the uncertainties about X and Z. The difference between $G(X)$ and Z is the 'error' introduced by the model G. We may refer to this 'error' as model inaccuracy or model uncertainty. It obviously needs to be addressed as the uncertainty assessments are conditional on the use of this model. The issue we raise here is how to deal with this 'error' – should we try to quantify it?

It would depend on the purpose of the analysis. Of course, when observations of Z are available, we would compare the assessments of Z, which are conditional on the use of the model G, with these observations. The result of such a comparison provides a basis for improving the model and accepting it for use. But at a certain stage we accept the model and apply it to support decision-making concerning, for example, risk acceptance and choice of options. Then it has no meaning in quantifying model uncertainty (inaccuracy). Instead of specifying $\mathrm{P}(Z \leq \mathrm{z})$ directly, we compute $\mathrm{P}(G(X) \leq z|K)$ and G is a part of the background knowledge K, or we choose to focus on summarizing measures such as the mean – that is, instead of specifying $\mathrm{E}Z$ we compute $\mathrm{E}[G(X)]$. In the following our main focus will be on the use of the model in such a context, to support decision-making concerning risk acceptance and the choice of options.

An important task for scientific communities in different areas is to develop good models. The models are justified by reference to established theories and laws explaining the phenomena studied, and the results of extensive testing. The performance of a model must, however, always be seen in light of the purpose of the analysis. A crude model can be preferred to a more accurate model in some situations if the model is simpler and it is able to identify the essential features of the system performance. Let us look at some more examples.

Example. Structural reliability analysis

In structural reliability analysis, attempts are often made to explicitly reflect the model uncertainties (inaccuracies); see, for example, Aven (2003). Let Z be the true capacity of the system at the time of interest. Using the model $G(X) = X_1 - X_2$, where X_1 represents a strength measurement and X_2 represents a load measurement, we have put $Z = G(X)$. This means a simplification, and the idea is then to introduce an error term X_0 (say), such that we obtain a new model $G_0(X) = X_0(X_1 - X_2)$. Clearly, this may give a better model, a more accurate description of the world. However, it may not be chosen in a practical case as it may complicate the assessments. It may be much more difficult to specify a probability distribution for (X_0, X_1, X_2) than for (X_1, X_2). There might be a lack of relevant data to support the uncertainty analysis of X_0 and there could be dependencies between X_0 and (X_1, X_2). We have to balance the need for accuracy and simplicity.

Example. Cost risk

A cost risk analysis is a tool typically used in project risk management (Aven, 2003). The purpose of the analysis is to assess the project cost and provide an evaluation of the uncertainties involved. To this end a model is developed linking the total cost of the project to a number of subcosts (cost elements), expressing costs related to different work packages. As an illustration we will consider a simple case where the total cost Z is written as a sum of two cost quantities, X_1 and X_2. Thus, we set $Z = G(X) = X_1 + X_2$.

A cost estimate (prediction) of Z is obtained by summing cost estimates (predictions) of X_1 and X_2. This is straightforward. In addition to the cost estimate (prediction), an uncertainty interval is normally produced. Assuming normal distributions, a 68% uncertainty interval is established by the cost estimate (prediction) $\pm$ the standard deviation σ, a 95% uncertainty interval by the cost estimate (prediction) $\pm 2\sigma$. If X_1 and X_2 are considered independent, this standard deviation is obtained from the standard deviation of X_i, denoted σ_i, $i = 1, 2$, by the formula

$$\sigma^2 = \sigma_1^2 + \sigma_2^2,$$

which is derived from the fact that the variance of a sum of independent random variables is equal to the sum of the variances of the random variables. If dependence is to be incorporated, the standard deviation σ is adjusted so that

$$\sigma^2 = \sigma_1^2 + \sigma_2^2 + 2\rho\sigma_1\sigma_2,$$

where ρ is the correlation coefficient between X_1 and X_2. Consider a case where the cost estimate is 5.0 for both X_1 and X_2, the standard deviations for X_1 and X_2 are 1.0 and 2.0 respectively, and the correlation coefficient is 0.5. Then a cost estimate (prediction) of 5.0 ± 2.6 is reported, with confidence about 70%. The cost estimates and standard deviations are established based on experience data whenever such data exist. Expert judgements are also used.

In many applications, the uncertainty is specified as relative values of the costs. Suppose that X_1 and X_2 are judged to be independent and the cost estimates (predictions) for X_1 and X_2 are 2 and 3, respectively. Furthermore, suppose that the uncertainty is $\pm 50\%$ relative to the costs, that is, ± 1 and ± 1.5 respectively, with a confidence of $3\sigma_i$, which is about 0.999. Then the cost estimate of 5 is presented with a reduced uncertainty of $\pm 36\%$, as 3σ is given by the square root of

$$(3\sigma_1)^2 + (3\sigma_2)^2 = 1^2 + 1.5^2 = 1.8^2,$$

which is 36% of 5.

In practice, Monte Carlo simulation is often used for the calculation of the uncertainty intervals. Using this technique, realizations of the system or activity (here the cost quantities) being analysed are generated, and based on these realizations the desired probability distributions can be established. When using Monte Carlo simulation, distributions other than the normal can easily be handled, such as triangular distributions, and complex dependence structures can be incorporated.

The model $G(X)$ is simple but should be sufficiently accurate for its purpose provided that all main cost elements have been included. If we could experience a cost element X_3 such that Z is poorly represented by $X_1 + X_2$, this cost element should, of course, be included in G. The problem is that such surprises could be difficult to anticipate before their occurrence. The analysts, being experts in the field of cost risk analysis, have studied data from other similar projects and carefully examined the possible scenarios that could lead to such surprises, and then they have to decide what is a sufficiently good representation of Z. The model is validated and accepted for use. The focus of this analysis is the model uncertainty (inaccuracy) $Z - G(X)$. Probabilities (knowledge-based) are not used to assess the model uncertainty (inaccuracy), but an empirical distribution may be established if data exist showing the difference between Z and $G(X)$. An example is shown in Figure 13.2. We see that if these data can be considered

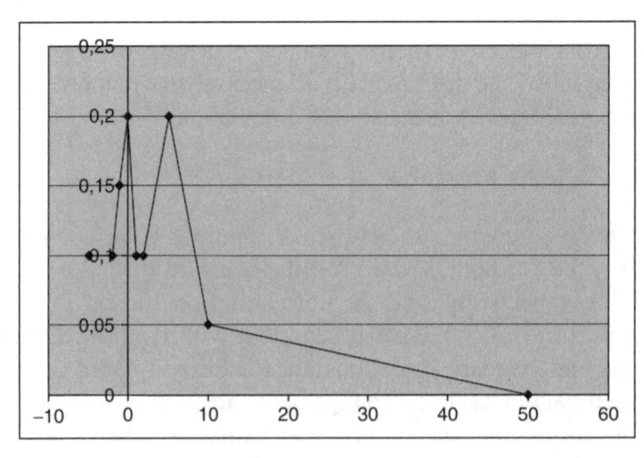

Figure 13.2 An empirical distribution for the model inaccuracy $Z - G(X)$ based on data from similar projects.

relevant, the model produces somewhat low values compared to the actual costs Z. It seems that some cost elements are ignored. Further analysis of the data revealed that two more cost elements should be considered to be included in the model. Using the extended model, the empirical distribution corresponding to Figure 13.2 showed considerably less variation. However, it was acknowledged that this analysis should be not given too much weight, as the relevance of the data could be questioned. The activity studied deviates from the observations on several issues. To obtain a sufficiently strong database, the population considered was extended to include many observations that could be seen as moderately relevant.

Now suppose that there is a rather strong dependence between the cost elements and that we would like to analyse this dependence. Correlation coefficients may be used to reflect the dependences, but this approach does not provide much insight about the factors causing the dependences. We may, for example, think of a situation where the cost elements are all strongly influenced by the oil price, and to estimate (predict) the overall cost Z explicit incorporation of this factor may be adequate. A simple way of doing this is shown in the following.

Let V be the value of an underlying factor, for example the oil price, influencing the cost elements X_i. It is common to refer to V as a latent quantity (variable). We write $X_i(V)$ to show the dependency of V. Given V, we judge the cost elements to be independent. Then, by specifying an uncertainty distribution of V, and of X_i given V, we can compute the uncertainty distribution of Z. By Monte Carlo simulation this is rather easy to do. We draw a number v from the distribution of V, and then use this as a starting point for drawing values of $X_i(v)$. We then use these data to produce a value of Z. The same procedure is repeated until we obtain the resulting probability distribution of Z.

The challenge is to find a simple way of expressing the judged dependencies. In the example above, where the X_i are related to the quantity V, we may go one step further and express X_i, for example, by the equation

$$X_i = a_i V + b_i + X_i', \tag{13.1}$$

where the quantities X_i' and V are judged independent, and X_i' has a distribution F_i', with mean 0 and standard deviation τ_i. By (13.1) the influence of the factor V on X_i has been explicitly described through remodelling, such that independence of the adjusted quantities $X_i - a_i V$ can be justified. It follows that

$$Z = \left(\sum a_i\right) V + \sum b_i + \sum X_i',$$

and this distribution can be found rather easily, for example by Monte Carlo simulation, as all unknown quantities on the right-hand side of the equality sign are judged independent. The basis for using equation (13.1) would normally be a regression analysis. The idea is to plot (using a so-called scatter plot – see Chapter 16) observations (v, x_i) of (V, X_i) on a two-dimensional diagram and fit the data to a line adopting standard linear regression.

Thus, we have obtained a new model $G'(X', V) = \left(\sum a_i\right) + \sum b_i + \sum X'_i$. This model is then assessed and possibly accepted for use.

Using normal distributions, it is sufficient for establishing the joint distribution of the X_i to specify the marginal distribution for each uncertain quantity X_i and the correlation coefficients of each pair of the X_i. Using some transformations of the marginal distributions, we can generalize this result. It is not straightforward, as we need to specify correlation coefficients of these transformations and not the correlation coefficients of X_i and X_j. See Bedford and Cooke (2001, p. 329).

An interesting alternative approach for specifying the joint distribution is presented in Bedford and Cooke (1999); see also Bedford and Cooke (2001). It is based on the specification of the marginal distributions, as well as probabilities of the form $\mathrm{P}(X_1 > x_1 | X_2 > x_2)$, where x_1 and x_2 are the 50% quantiles of the distributions of X_1 and X_2, respectively. Using a mathematical procedure, a minimal informative distribution is established based on this input. A minimal informative distribution is, in a sense, the most 'independent' joint distribution with the required properties.

Reflection

The above probabilities are knowledge-based probabilities, expressing the analysts' uncertainty about unknown quantities. Could we alternatively have used frequentist probabilities P_f?

If we were to adopt frequentist probabilities instead of knowledge-based probabilities, the above probabilities would have to be replaced by estimates. For example, instead of estimating $P_\mathrm{f}(Z \leq z)$ directly, we estimate $P_\mathrm{f}(G(X) \leq z)$. Uncertainties of the P_f also need to be addressed and by expressing these by means of knowledge-based probabilities, we are back in the set-up introduced above with the P_f considered as unknown quantities X and Z. See the examples below (on lifetime distributions and the parallel system) and also Chapters 12 and 20.

Example. Dropped object

A simple physical model is the expression for the speed v of an object dropped from a height h, derived by assuming that the kinetic energy of the object at the reference point equals the potential energy at h (Nilsen and Aven, 2003):

$$v = \sqrt{2gh}, \tag{13.2}$$

where g is the acceleration due to gravity.

Now consider using (13.2) to predict the velocity of an object dropped from a crane located on a floating structure. It can be argued that the vertical motion of the structure due to ocean waves would cause the model to be inaccurate. Such motion would affect the height of fall h, cause an initial speed $v_0 \neq 0$, and a relative vertical motion of the object hit by the dropped object – three effects that are not taken into account by the model. Another phenomenon not considered by (13.2) that would cause non-compliance for any dropped object is air resistance, which is influenced by the mass and the shape of the object – quantities that are not included in (13.2).

Nonetheless, the model may be accepted as it provides insights and is a simple tool for predicting the velocity of the dropped object. Depending on the purpose of the analysis, the model could be judged sufficiently accurate. Tests can be carried out to demonstrate the 'goodness' of the model. If these tests show sufficiently accurate results, the model is accepted and is used in the risk analysis to assess the risk associated with dropped objects.

Example. Lifetime distributions

Consider the lifetime distribution $F(t)$ of a mass-produced unit which expresses the proportion of units with lifetime equal to or less than t. The exponential distribution $G(t|\lambda)$ is a model of F, where λ is the parameter of the distribution, the failure rate. Model inaccuracy expresses the difference between this model and the true distribution. The parameter λ we interpret as the inverse of the average lifetime in the infinite population of the units. Epistemic uncertainties about λ are incorporated by using the law of total probability:

$$P(T \leq t) = \int G(t|\lambda)dH(\lambda), \tag{13.3}$$

where T is the lifetime and H is the uncertainty distribution of λ. Remember that P is a knowledge-based probability. This probability is used as an input in a reliability analysis of the system, to support the decision-making on risk acceptance and choice of options. Model uncertainty related to the choice of probability distribution is not reflected. We have specified $\mathrm{P}(T \leq t)$ using the exponential distribution according to formula (13.3); the probability is conditional on the use of this distribution. If we knew the true distribution, P would be different, but since this distribution is unknown we condition on this model. We have introduced the exponential distribution to simplify the problem. If we had considered the space of all distribution functions, the assignment process would not be feasible in practice. If we are not satisfied with the exponential distribution class, we should change this by using, for example, a Weibull distribution. But when performing the analysis to support the decision-making on risk acceptance and choice of options, and computing $\mathrm{P}(T \leq t)$, we accept the use of a specific model, which constitutes a part of the background knowledge that the assessment is based on.

Example Continued. Parallel System

Let us return to the analysis of the parallel system comprising two components, as shown in Figure 13.1. Now we consider a problem closely related to the one first studied. Let us define the model by

$$G(p) = 1 - (1 - p_1)(1 - p_2),$$

where p_i is understood as the proportion of functioning components when considering an infinite number of components similar to component i. We refer to p_i as a 'chance'. The quantity of interest in this case is another chance, the system reliability h, defined as the proportion of systems functioning when considering an infinite number of such systems. The set-up is as in Lindley and Singpurwalla (2002). By expressing knowledge-based probabilities P about the chances p_i we obtain an uncertainty distribution about h. The analysis may be carried out using Monte Carlo simulation. The analysis is an example of a typical uncertainty analysis in a reliability and risk context, a probability of frequency analysis (see Chapter 2).

As described above in relation to (13.3), a model may be introduced to express p_i, for example the exponential model, leading to

$$p_i = \exp\{-\lambda_i t\},$$

where λ_i is the failure rate and t is a fixed point in time. Based on a probability distribution for λ_i we obtain distributions for p_i and h. Following a standard Bayesian statistical analysis, we first specify a prior distribution for λ_i and then use the Bayesian updating procedure to obtain the posterior distribution when data become available (see Chapter 15).

If the state of the system, Z, is the quantity of interest, we may first establish a distribution of the state of the components (X_1, X_2) by expressions such as

$$\mathrm{P}(X_1 = 0 \text{ and } X_2 = 0) = \int (1 - \exp\{-\lambda_1 t\})(1 - \exp\{-\lambda_2 t\}) dH(\lambda),$$

where H is the uncertainty distribution of $\lambda = (\lambda_1, \lambda_2)$. Then, using the model $G(X) = 1 - (1 - X_1)(1 - X_2)$, we can compute $\mathrm{P}(Z = 1)$. We thus carry out a two-stage uncertainty analysis, firstly for establishing the distribution of (X_1, X_2). In this case the exponential distribution is the model of the set-up and λ is the input quantity. In the second run, $G(X) = 1 - (1 - X_1)(1 - X_2)$ is the model and X is the input.

In the literature various methods have been suggested to reflect model uncertainties (Apostolakis, 1990; Nilsen and Aven, 2003; Devooght, 1998; Zio and Apostolakis, 1996). Let us look at one basic approach (Apostolakis, 1990) which addresses the issue of weighting different models. Let M_1 and M_2 be two alternative models to be used for assigning the probability A. Conditional on M_i, we have an assignment $\mathrm{P}(A|K_i)$. Unconditionally, this gives

$$\mathrm{P}(A|K) = \mathrm{P}(A|K_1)p_1 + \mathrm{P}(A|K_2)p_2, \tag{13.4}$$

where p_i is the analyst's subjective probability that the ith model (i.e. the set of associated assumptions) is true.

Such a procedure is analogous to the exponential lifetime case (13.3) if 'true' refers to a condition of the real world that is true or not. We have introduced the model to simplify a complex world, and it is meaningless to talk about whether the model is true or not. In a practical decision-making context the analysts would most likely present separate assignments for the different models $\mathrm{P}(A|K_i)$, in addition to the weighed probability assignment (13.4). To specify the subjective probability $\mathrm{P}(A|K)$ the analysts may choose to apply the assignment procedure given by (13.4) also when p_i cannot be interpreted as a probability that a specific assumption is true. In such a case p_i must be interpreted as a weight reflecting the confidence in the model i for making accurate predictions.

There will always be a need for clarification of what is uncertain and subject to the uncertainty assessment and what is to be considered fixed. It is not possible to perform a risk analysis without making assumptions. The results of the analysis always have to be seen in light of these assumptions.

Models introduce parameters, as in the above examples. These parameters need to have a clear interpretation as expressing uncertainties about a poorly defined quantity is meaningless. There is a tendency in many applications to introduce chances and frequentist probabilities, also when such constructions are based on fictional populations. Then the interpretation is not always clear. Consider the example in Chapter 12 of the frequentist probability (chance) p of at least one fatality in 1 year in a process plant. According to the relative frequency view, this probability is interpreted as the proportion of plants with at least one fatality when considering an infinite number of similar plants. But as discussed in Chapter 12, we will have a hard time explaining the meaning of p. How can we then perform a meaningful uncertainty analysis? The point we are making is that for the assessment to make sense, all quantities and parameters introduced must be properly defined.

Summary

In this chapter we have discussed the issue of model uncertainty (in the sense of model inaccuracy). Should we quantify this uncertainty?

Our clear answer is no. It is not meaningful to quantify the model uncertainty (inaccuracy). Our point is that if the model is not considered good enough for its purpose, it should be improved. The uncertainty assessments are based (conditional) on the model used. However, to test and validate a model, we will, of course, address the accuracy of the model, but when using it for supporting decisions on, for example, risk acceptance and choice of options, we need to base our assessment on a specific model. At this stage quantifying the model uncertainty (inaccuracy) has no meaning. If we could measure the model uncertainty (inaccuracy), there would be no need for the model.

References

Apostolakis, G. (1990) The concept of probability in safety assessments of technological systems. *Science*, **250**, 1359–1364.

Aven, T. (2003) *Foundations of Risk Analysis*, John Wiley & Sons, Ltd, Chichester.

Bedford, T. and Cooke, R. (1999) A new generic model for applying MAUT. *European Journal of Operational Research*, **118**, 589–604.

Bedford, T. and Cooke, R. (2001) *Probabilistic Risk Analysis. Foundations and Methods*, Cambridge University Press, Cambridge.

Devooght, J. (1998) Model uncertainty and model inaccuracy. *Reliability Engineering and System Safety*, **59**, 171–185.

Lindley, D.V. and Singpurwalla, N. (2002) On exchangeable, causal and cascading failures. *Statistical Science*, **17** (2), 209–219.

Nilsen, T. and Aven, T. (2003) Models and model uncertainty in the context of risk analysis. *Reliability Engineering and System Safety*, **79**, 309–317.

Zio, E. and Apostolakis, G.E. (1996) Two methods for the structured assessment of model uncertainty by experts in performance assessments of radioactive waste repositories. *Reliability Engineering and System Safety*, **54**, 225–241.

Further reading

Aven, T. (2009) Some reflections on uncertainty analysis and management. Accepted for publication in *Reliability Engineering and System Safety*.

14

It is meaningful and useful to distinguish between stochastic and epistemic uncertainties

Let us again return to the die ('Russian roulette') game introduced in Chapter 1, where John is offering you one game with the following payoffs: if the die shows a 6 you lose \$24 000, otherwise you win \$6000. John selects the die but you throw it. To assess the uncertainties, the prevailing approach among risk analysts would be to distinguish between (Paté-Cornell, 1996):

1. **Aleatory uncertainties** (often also referred to as 'stochastic uncertainties' and 'randomness'): the variation in outcomes or populations (here when throwing the die).

2. **Epistemic uncertainties:** the lack of knowledge about fundamental phenomena, that is, about the true outcome distribution.

Aleatory uncertainties arise because the system under study (in our case the die) can behave in many different ways and these uncertainties are thus a property of the system. Epistemic uncertainties, which arise from a lack of knowledge about the system, are on the other hand a property of the analysts performing the study (Helton, 1994). Epistemic uncertainties are reducible by obtaining more or better information, whereas aleatory uncertainties are irreducible.

This is the way these concepts are commonly explained in the literature. However, a further look into the concepts reveals that their meaning is not so clear. Many researchers have pointed to the fact that uncertainty is uncertainty, and attempting to distinguish between 'types of uncertainty' is questionable (Winkler, 1996). There is only one kind of uncertainty and this uncertainty stems from our lack of knowledge concerning the truth of a proposition or the value of an unknown quantity. Winkler (1996) uses a coin example to illustrate the difficulty of separating aleatory and epistemic uncertainties:

> Consider the tossing of a coin. If we all agree that the coin is fair, then we would agree that the probability that it lands heads the next time it is tossed is one-half. At first glance, our uncertainty about how it lands might be thought of as aleatory, or irreducible. Suppose, however, that the set of available information changes. In principle, if we knew all of the conditions surrounding the toss (the initial side facing up; the height, initial velocity, and angle of the coin; the wind; the nature of the surface on which the coin will land; and so on), we could use the laws of physics to predict with certainty or near certainty whether the coin will land heads or tails. Thus, in principle, the uncertainty is not irreducible, but is a function of the state of knowledge (and hence is epistemic).
>
> In practice, of course, measuring all of the initial conditions and doing the modelling in the coin-tossing example are difficult and costly at best and would generally be viewed as infeasible. (However, if a billion dollars was riding on the toss of the coin, we might give serious consideration to some modelling and measurement!) Our uncertainty about all of the initial conditions and our unwillingness to spend time to build a detailed model to relate the initial conditions to the ultimate outcome of the toss translate into a probability of one-half for heads. Even uncertainty about just the initial side that faces up could translate into a probability of one-half for heads. At a foundational level, as noted above, our uncertainty in a given situation is a function of the information that is available. (Winkler, 1996)

Winkler (1996) concludes, as do many others, including Lindley (2006) and Aven (2003), that all uncertainties are epistemic uncertainties, a result of lack of knowledge. However, for the purpose of analysing uncertainties and risk it may be useful to introduce models – and aleatory uncertainty represents a way of modelling the phenomena studied. In the case of a coin, the model is that if we toss the coin over and over again, the fraction of heads will be p. Similarly, for the die example, we would establish a model expressing that the distribution of outcomes is given by $(p_1, p_2, \ldots, p_6)$, where p_i is the fraction of outcomes showing i. These fractions are parameters of the models, and they are referred to as probabilities in a traditional classical statistical setting and as chances in the Bayesian setting.

The models are often referred to as probability models or stochastic models. They constitute the basis for statistical analysis, and are considered essential for assessing the uncertainties and drawing useful insights (Winkler, 1996; Helton,

1994). The probability models coherently and mechanically facilitate the updating of probabilities. All analysts and researchers acknowledge the need to decompose the problem in a reasonable way, but many would avoid the reference to different types of uncertainties as they consider all uncertainties to be epistemic.

The traditional statistical analysis restricts attention to frequentist probabilities, whereas the Bayesian approach allows incorporation of all available evidence in the assessments. The Bayesian approach introduces unknown chances which are subject to epistemic uncertainties, and the result is a measure of uncertainty reflecting both variation in populations and the epistemic uncertainties about the true value of the chances.

If we study the lifetimes of a type of light bulb, the distinction between variation (aleatory uncertainty) and epistemic uncertainty is clear and easy to understand. The different lifetimes produces the variation. If we had full knowledge about the generated distribution of lifetimes, there would be no epistemic uncertainties. However, in practice, we do not know this underlying true distribution and hence we need to address epistemic uncertainties. In practice, this is typically done by assuming that the distribution belongs to a parametric distribution class, for example the Weibull distribution with parameters α and β. This distribution is to be considered a model of the underlying true distribution, and given this model the epistemic uncertainties are reduced to lack of knowledge concerning the correct parameters α and β.

For mass-produced units and other situations with large populations of units, this uncertainty structure makes sense. In line with Winkler (1996) we would, however, prefer to refer to the aleatory uncertainty as variation and not uncertainty, as variation or population variation better explains the meaning of the concept.

In a risk assessment context, the situations are often unique, and the distinction between variation (aleatory uncertainty) and epistemic uncertainty is then more problematic. Consider as an example the probability of a terrorist attack (properly specified). To define the aleatory uncertainty in this case we need to construct an infinite population of similar attack situations. The variation in this population generated by 'success' (attack) and 'failure' (no attack) represents the aleatory uncertainty. The proportion of successes equals the probability of an attack. But is such a construction meaningful? No – it makes no sense to define a large set of 'identical', independent attack situations, where some aspects (e.g. related to the potential attackers and the political context) are fixed and others (e.g. the attackers' motivation) are subject to variation. Suppose that the attack success rate is 10%. Then, in 1000 situations, with the attackers and the political context specified, the attackers will attack in about 100 cases. In these situations the attackers are motivated, but not in the remaining ones. Motivation for an attack in one situation does not affect the motivation in another. For independent random situations (refer to the light bulb example above) such 'experiments' are meaningful, but not in unique cases like this.

This type of problem is addressed surprisingly seldom in the literature. It is common to introduce the underlying aleatory-based probabilities and distributions, but without clarifying their meaning. Researchers who take the view that there is only one type of uncertainty, stemming from lack of knowledge (i.e. it is epistemic), seem to represent a small minority; see, for example, Helton and Burmaster (1996). Yet, the Bayesian paradigm, as presented for example by Lindley (2006), is based on this idea. Probability is considered a measure of uncertainty about events and outcomes (consequences), seen through the eyes of the assessor and based on the available background information and knowledge. Probability is a subjective measure of uncertainty, conditional on the background knowledge. The reference is a certain standard such as drawing a ball from an urn. If we assign a probability of 0.4 to an event A, we compare our belief about the occurrence of A with drawing a red ball from an urn having 10 balls where 4 are red.

According to this paradigm, there is epistemic uncertainty associated with the occurrence of a terrorist attack. And the analyst assigns a probability expressing his/her uncertainty about this event. If relevant, knowledge about variability is included in the background knowledge. The variability gives rise to uncertainty but is not defined as uncertainty in this context.

Summary

All uncertainties are epistemic, a result of lack of knowledge. However, for the purpose of analysing uncertainties and risk it may be useful to introduce models based on variations in populations. This variation is often referred to as aleatory uncertainty, although it does not represent uncertainty for the assessor.

References

Aven, T. (2003) *Foundations of Risk Analysis*, John Wiley & Sons, Ltd, Chichester.

Helton, J.C. (1994) Treatment of uncertainty in performance assessments for complex systems. *Risk Analysis*, **14**, 483–511.

Helton, J.C. and Burmaster, D.E. (eds) (1996) Special issue on treatment of aleatory and epistemic uncertainty. *Reliability Engineering and System Safety*, **54** (2–3).

Lindley, D.V. (2006) *Understanding Uncertainty*, John Wiley & Sons, Inc., Hoboken, NJ.

Paté-Cornell, M.E. (1996) Uncertainties in risk analysis: six levels of treatment. *Reliability Engineering and System Safety*, **54** (2–3), 95–111.

Winkler, R.L. (1996) Uncertainty in probabilistic risk assessment. *Reliability Engineering and System Safety*, **54** (2–3), 127–132.

Further reading

Aven, T. (2009) On the need for restricting the probabilistic analysis in risk assessments to variability. Accepted for publication in *Risk Analysis*.

15

Bayesian analysis is based on the use of probability models and Bayesian updating

We consider again the die game studied in previous chapters, where John is offering you one game with the following payoffs: if the die shows a 6, you lose \$24 000, otherwise you win \$6000. Johns selects the die but you throw the die.

Now suppose that John is offering you the possibility of throwing the die once. How should you incorporate this information in your assessment?

A well-established approach is to use the Bayesian updating procedure. We establish a stochastic model of the situation analysed. Let p_i be the fraction of tosses of the die resulting in the outcome i if the experiment is repeated infinitely, $i = 1, 2, \ldots, 6$. The fractions p_i are referred to as chances and $(p_1, p_2, \ldots, p_6)$ as the chance distribution of the outcome. To apply the Bayesian procedure you first have to specify a prior distribution for p_6, which expresses your initial uncertainty (degree of belief) about the true value of p_6. Suppose you assign the distribution for p (for the sake of simplicity we write p instead of p_6) as shown in Table 15.1. The probabilities here are knowledge-based probabilities.

Hence, you presume that p could either be $1/6$ (a normal, fair die), 0.5, 0.9 or 1. Now let X denote the number of 6s when tossing the die, that is, X equals 1 if the die shows 6, and 0 otherwise, since the die is thrown only once.

Let us assume that you throw the die and you observe $X = 0$, that is, the die does not show 6. What would then be your uncertainties (degree of beliefs) about p? The result is shown in Table 15.2. This probability distribution, which is referred to as the posterior distribution, is computed using Bayes' formula, according to which $\mathrm{P}(A|B) = \mathrm{P}(B|A)\mathrm{P}(A)/\mathrm{P}(B)$, for events A and B.

Misconceptions of Risk T. Aven

Table 15.1 Prior distribution for p.

p'	$\mathrm{P}(p = p')$
1	$1/4$
0.90	$1/4$
0.50	$1/4$
1/6	$1/4$

Table 15.2 Posterior distribution for p.

p'	$\mathrm{P}(p = p' \mid X = 0)$
1	0
0.90	0.07
0.50	0.35
1/6	0.58

To show the calculations, let $d = 1/\mathrm{P}(X = 0)$. Then

$$\mathrm{P}(p = 1/6|X = 0) = \mathrm{P}(X = 0|p = 1/6)\ \mathrm{P}(p = 1/6)\ d = d \cdot \tfrac{5}{6} \cdot \tfrac{1}{4} = 0.208d.$$

Furthermore,

$$\mathrm{P}(p = 0.5|X = 0) = \mathrm{P}(X = 0|p = 0.5)\ \mathrm{P}(p = 0.5)\ d = d \cdot 0.5 \cdot \tfrac{1}{4} = 0.125d,$$

$$\mathrm{P}(p = 0.9|X = 0) = \mathrm{P}(X = 0|p = 0.9)\ \mathrm{P}(p = 0.9)\ d = d \cdot 0.1 \cdot \tfrac{1}{4} = 0.025d,$$

$$\mathrm{P}(p = 1|X = 0) = \mathrm{P}(X = 0|p = 1)\ \mathrm{P}(p = 1)\ d = d \cdot 0.0 \cdot \tfrac{1}{4} = 0.$$

Obviously, if $X = 0$, we cannot have $p = 1$. Hence, $0.208d + 0.125d + 0.025d = 1$, and $d = 2.79$. The distribution in Table 15.2 follows.

We see that initially your probability of the die being normal was 25%. After observing $X = 0$, this probability has increased to approximately 60%. Perhaps John is not cheating?

Let us assume that John is offering you the possibility to toss the die one more time, and let us assume that the second throw also shows an outcome different from 6, that is, $X = 0$. What would then be the posterior distribution? The result is shown in Table 15.3. The computations are analogous to those for one throw. We simply repeat the above calculations using the distribution of Table 15.1 as the new prior distribution. This gives

$$\mathrm{P}(p = 1/6|X = 0) = \mathrm{P}(X = 0|p = 1/6)\ \mathrm{P}(p = 1/6)\ d = d \cdot \tfrac{5}{6} \cdot 0.58 = 0.483d,$$

$$\mathrm{P}(p = 0.5|X = 0) = \mathrm{P}(X = 0|p = 0.5)\ \mathrm{P}(p = 0.5)\ d = d \cdot 0.5 \cdot 0.35 = 0.175d,$$

$$\mathrm{P}(p = 0.9|X = 0) = \mathrm{P}(X = 0|p = 0.9)\ \mathrm{P}(p = 0.9)\ d = d \cdot 0.1 \cdot 0.07 = 0.007d,$$

$$\mathrm{P}(p = 1|X = 0) = 0.$$

Hence $0.483d + 0.175d + 0.007d = 1$ which gives $d = 1.5$ and the distribution in Table 15.3 is established. We see that the probability that the die is fair is now increased to 73%.

Table 15.3 Posterior distribution for p when throwing the die twice.

p'	$\mathrm{P}(p = p'\|X = 0)$
1	0
0.90	0.01
0.50	0.26
1/6	0.73

Reflection

In the above analysis the Bayes procedure was used twice. Would the results be the same if we performed just one run of the Bayesian procedure, starting from the prior distribution of Table 15.1 and then incorporating the information from observing two throws both showing outcomes different from 6?

The calculations and results will be the same. We have observed $X = 0$, meaning that both throws are successes in the sense that the outcome 6 is avoided. We obtain

$$\mathrm{P}(p = 1/6|X = 0) = \mathrm{P}(X = 0|p = 1/6)\ \mathrm{P}(p = 1/6)\ d = d \cdot (5/6)^2 \cdot \tfrac{1}{4} = 0.173d,$$

$$\mathrm{P}(p = 0.5|X = 0) = \mathrm{P}(X = 0|p = 0.5)\ \mathrm{P}(p = 0.5)\ d = d \cdot 0.5^2 \cdot \tfrac{1}{4} = 0.0625d,$$

$$\mathrm{P}(p = 0.9|X = 0) = \mathrm{P}(X = 0|p = 0.9)\ \mathrm{P}(p = 0.9)\ d = d \cdot 0.1^2 \cdot \tfrac{1}{4} = 0.0025d,$$

$$\mathrm{P}(p = 1|X = 0) = 0.$$

This gives the same posterior distribution as above. □

Reflection

What is your probability that the next (third) throw shows an outcome 6, given that the first two throws have not shown the outcome 6?

This probability we can compute using the law of total probability (all probabilities are conditioned on the event that the first two throws have not shown the outcome 6):

$$\begin{aligned}\mathrm{P}(6) &= P(6|p = 1/6)\ \mathrm{P}(p = 1/6) + \mathrm{P}(6|p = 0.5)\ \mathrm{P}(p = 0.5)\\ &\quad + \mathrm{P}(6|p = 0.9)\ \mathrm{P}(p = 0.9)\\ &= \tfrac{1}{6} \cdot 0.73 + 0.5 \cdot 0.26 + 0.90 \cdot 0.01 = 0.26.\end{aligned}$$

Hence P(not 6) $= 0.74$. We see that P(6) equals the expected value of p, that is, $p(6) = \mathrm{E}[p]$.

The probability distribution (P(6), 1 – P(6)) is referred to as the predictive distribution and integrates the posterior distribution and the chance distribution. Using the terminology of the previous chapter, the predictive distribution integrates epistemic uncertainties expressed by the posterior distribution, and the aleatory uncertainty (variation) expressed by the chances $P(6|p) = p$. □

Reflection

What information is of most relevance for you when deciding whether you should accept the game or not: the posterior distribution of p expressed in Table 15.2, or the predictive distribution (P(6), 1 – P(6))?

Both distributions provide useful decision support, but P(6) is the most interesting probability as it directly expresses your probability that the outcome is a loss of $24 000. The posterior distribution relates to behaviour of the die in the long run, but this game is based on one throw only (following the two first test throws). □

This Bayesian approach is a strong machine for updating probabilities when new information becomes available. From the above analysis, we see how the machinery works. First establish a probabilistic model, then assign a prior distribution on the parameter of interest. Next use Bayes' theorem to establish the posterior distribution, and finally compute the predictive distribution using the law of total probability. This machinery has a solid theoretical basis, but is not always easy to use in practice. The first problem is to define the set of possible parameter values and assign a prior distribution. In our case, we chose to use the values 1/6, 0.50, 0.90 and 1.0. But why exactly these values? And what is the justification for using the prior probability numbers ¼ for each of these four possible values of p? Using the Bayesian machinery we obtain mathematical rigour, but would the produced numbers be of less strength if you simply expressed your knowledge-based probability based on all relevant information? You look at John, and the results of the tests, and then you assign a probability $P(6) = 0.30$ (say) based on all the available background knowledge. No probabilistic model is introduced, and hence there is no need for a justification of the prior distribution. For this particular case such an approach could obviously be sufficient, in particular for the case when no tests are performed, but in general it would suffer from some serious weaknesses: it would be difficult to perform consistent analyses – the approach would lack rigour.

We will discuss these issues in more detail using some more realistic cases.

Bayesian updating for a drilling operation

Consider a well that is to be drilled from an offshore location and into a reservoir below the seabed (see Sandøy and Aven, 2006). The drilling operation is regarded as difficult because of large uncertainty about the pore pressure in the reservoir. If the well pressure drops below the formation pore pressure during drilling in the reservoir, a blowout may result, threatening human lives and the environment. On the other hand, too high a well pressure may result in a low drilling speed, differential sticking problems and, in the worst case, fracturing of the formation.

A risk analysis is performed as a part of the planning phase of the well. One of the objectives of this risk analysis is to identify an 'optimal' well pressure for drilling in the reservoir. The risk analysis is based on detailed modelling of mechanisms causing severe events, including geological, human and equipment aspects, and this detailed modelling allows analysis of the risk-reducing or risk-increasing effect of different decisions such as adjustments of the target well pressure.

However, in the planning phase when the risk analysis is performed, the uncertainty about some of the quantities included in the analysis is large. At the start of the drilling and during the operation, information is acquired that reduces the uncertainty about these quantities. The new information can be judged to be of such significance that decisions made in the planning phase are reconsidered, even though this process is not normally supported by a formal risk analysis.

Reconsidering decisions made on the basis of planning phase risk analyses, in light of new information made available during a drilling operation, can reduce the risk associated with the operation. But making decisions during a drilling operation, without support from a formal risk analysis, also represents a risk as optimization involving a large number of quantities is difficult to perform mentally, and especially under time pressure. Ideally, a manual update of the planning phase risk analysis should be performed, but the need for fast reaction during drilling operations does not allow for this.

To include the information acquired during the drilling operation in the basis for the decision, and to ensure that the implications of each decision are thoroughly analysed, a risk analysis tool is needed that automatically updates the planning phase risk analysis during the drilling operation as new information arrives. Such a tool would at all times reflect the available information and would be an important instrument for supporting decisions during the drilling operation. A procedure for updating assessments of quantities included in the risk analysis would be a vital part of such a risk analysis tool. To ensure that the risk analysis at all times reflects the available information, the procedure would need to be performed automatically, rapidly and without requiring input from personnel.

Below we outline how such a procedure can be established using Bayes' theorem. Our focus is the formation of pore pressure and the procedure for automatic updating of pore pressure assessments. Alternative approaches will be discussed.

Updating procedures for pore pressure assessments

Let θ be the real pore pressure at depth d, the depth in the reservoir where the difference between the pore pressure and the well pressure is expected to be at its minimum. If the well pressure is above the pore pressure at depth d, the well pressure will also be above the pore pressure at all other depths.

The value of θ is unknown, but information exists that gives an indication of what the real value is. The amount of information about θ increases with time as the planning and the drilling of the well proceed.

In the planning phase of the well, information about the pore pressure is in the form of seismic data and data from possible nearby wells. Let H represent everything we (the analysts) know about θ in the planning phase of the well.

During the drilling operation, observations that give information about the pore pressure θ are obtained. The value of the information provided by these observations depends on the depth at which the observations are performed, and the 'measurement equipment' used for making the observations.

Observations performed at the depth d are direct measurements of the quantity θ, and give valuable information about θ. However, observations performed at a lower depth might also give some information about θ. As a simplification, only observations performed when the well is believed to have reached depth d are considered. The expression 'believed to have reached depth d' reflects that both the depth d and the actual drilled depth are unknown quantities, and hence one cannot be certain about when the well has reached the depth d.

Some types of measurement equipment are considered to give more reliable results than others, and as a consequence the value of the information provided by the observations also depends on the measurement equipment on which the observations are made. In this presentation we consider observations from a single type of measurement equipment.

Let $x = (x_1, x_2, \ldots, x_n)$ represent the observations of the pore pressure θ, performed by use of a single unit of measurement equipment, when the well is believed drilled down to the critical depth d. In the following we present a Bayesian method for assessing the uncertainty about the real value of θ in the planning phase and updating the assessment of θ automatically in light of the observations x made during the drilling operation.

To start simply, consider a scenario where an assessment of θ is to be made in the planning phase of the well, and this assessment is to be updated based on the single observation x_1. First we assign a prior probability distribution F for θ such that $F(\theta'|H) = \mathrm{P}(\theta \leq \theta'|H)$. We assume that this function can be written as an integral of a density $f(\theta|H)$. In our case a normal distribution is used to express the uncertainties about θ, that is,

$$f(\theta|H) = \frac{1}{\sigma\sqrt{2\pi}} \exp\left\{-\frac{1}{2}\left(\frac{\theta-\mu}{\sigma}\right)^2\right\},$$

where μ and σ are the parameters of the distribution, the expected value and standard deviation, respectively. Based on the information available in the planning phase, H, experts assign the values $\mu = 1.71$ and $\sigma = 0.09$. The prior density is presented in Figure 15.1.

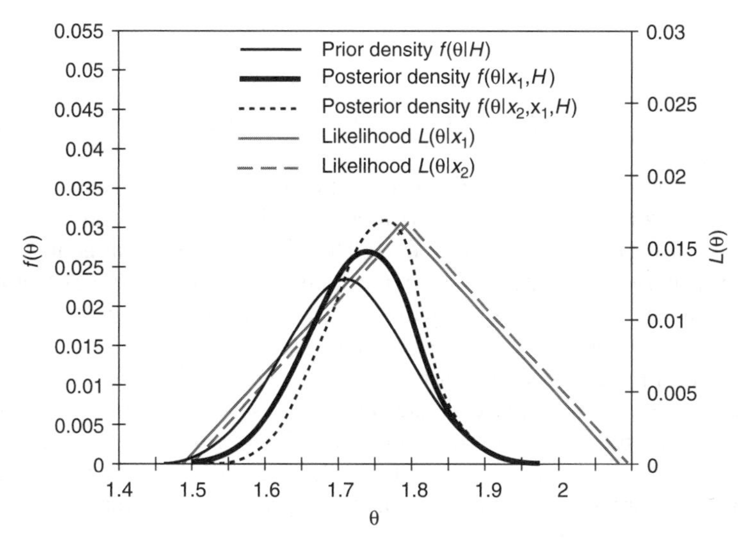

Figure 15.1 Uncertainty about θ: prior density, posterior density after first observation, and posterior density after second observation. The likelihood functions are plotted on the second axis (Sandøy and Aven, 2006). With permission of International Journal of Reliability, Quality and Safety Engineering World Scientific.

Next we assess the uncertainties about the accuracy of the measurement equipment. If θ is the true pressure we would expect the observation x_1 to be close to θ, but rather large deviations could also occur. This is reflected in the probability density of x_1 given θ, $f(x_1|\theta)$; see Figure 15.2. The density is triangular, with expected value at θ and all mass within the interval $(\theta - 0.3, \theta + 0.3)$.

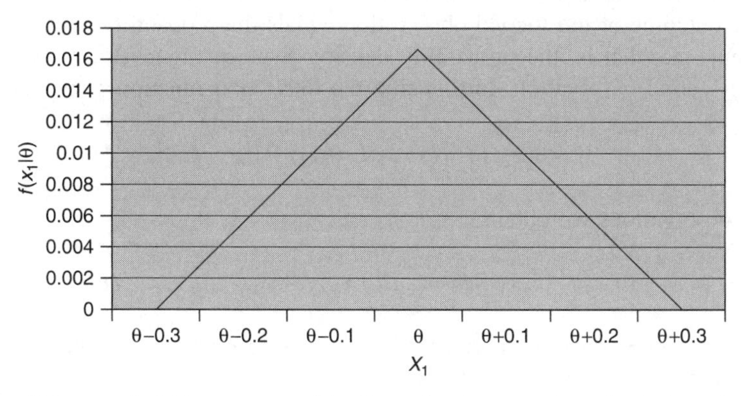

Figure 15.2 Probability density $f(x_1|\theta)$ reflecting possible deviations between the observation x_1 and the pore pressure θ (Sandøy and Aven, 2006). With permission of International Journal of Reliability, Quality and Safety Engineering World Scientific.

Bayes' theorem is then used to establish the posterior density of θ given the observation x_1:

$$f(\theta|x_1, H) = cf(x_1|\theta)f(\theta|H), \tag{15.1}$$

where c is a normalizing constant such that the integral over the density equals 1.

As a function of θ, $f(x_1|\theta)$ is written as $L(\theta|x_1)$ and is referred to as the likelihood of θ given the observation x_1. The likelihood function is not a probability distribution. It is interpreted as a scale of comparative support lent by the known x_1 to the various possible values of the unknown θ (Singpurwalla, 2006).

As the real value of θ is unknown, the probability density $f(x_1|\theta)$ should be assigned for all possible values of θ. However, as a simplification, the measurement equipment used to perform the observations is considered to be equally reliable for all pressures, and only one probability density reflecting possible deviations between the observation x_1 and the real value of θ is necessary. The probability density has its maximum when x_1 equals θ, and decreases for increasing deviations from this value.

To extend the problem somewhat, consider a scenario where another observation, x_2, is made. The prior density is now $f(\theta|x_1, H)$ and this gives the posterior density

$$f(\theta|x_2, x_1, H) = cf(x_2|\theta, x_1)f(\theta|x_1, H) = cf(x_2|\theta)f(\theta|x_1, H). \tag{15.2}$$

The second equality utilizes the fact that x_1 and x_2 are judged independent when conditioned on θ, that is, $f(x_2|x_1, \theta) = f(x_2|\theta)$. When θ is given, the observation x_1 has no influence on our beliefs/knowledge about the outcome of x_2 and the outcome of x_2 depends merely on the specific value of θ and the reliability of the measurement equipment. As only one type of measurement equipment and only deviations between the observation and θ are considered, the probability density for the outcome of the second observation equals the probability density for the first outcome, that is, the probability density given in Figure 15.2 is valid also for the second observation. Introducing the likelihood function L, we can write (15.2) as

$$f(\theta|x_2, x_1, H) = cL(\theta|x_2)f(\theta|x_1, H).$$

For n observations we obtain

$$f(\theta|x_n, \ldots, x_1, H) = cL(\theta|x_n)f(\theta|x_{n-1}, \ldots, x_1, H).$$

We observe $x_2 = 1.79$; see Figure 15.1. The first observation indicates a pressure that is higher than expected and, as can be seen from Figure 15.1, the mean value of the posterior density is increased based on this observation. The second observation is also higher than expected and this results in another increase in the mean. Note also that the variance is decreased when the two observations are included in the probability density reflecting the uncertainty about θ.

An alternative Bayesian updating approach

The objective of the updating procedure is to assess the uncertainty about the real value of θ. As a way to abridge the information H, a new unknown quantity ϕ, scalar or vector, is introduced. By the law of total probability and by judging θ independent of H if ϕ were to be known, we can write

$$f(\theta|H) = \int_{\phi} f(\theta|\phi) f(\phi|H) d\phi, \tag{15.3}$$

where

- $f(\theta|\phi)$ is a probability model for θ with parameter ϕ;
- $f(\phi|H)$ is the prior density reflecting the possible values that the parameter ϕ can take, based on the information H.

The parametric probability model $f(\theta|\phi)$ is introduced because the quantity θ is found difficult to assess directly. Introducing the probability model $f(\theta hh|\phi)$ allows assessment of the parameter ϕ instead of θ, and it may simplify the analysis as ϕ is presumed to be easier to assess compared to θ.

The interpretation of the parameter ϕ depends on the probability model chosen and the situation to be analysed, and can be an observable quantity reflecting a state of the nature, or an unobservable fictional quantity. To illustrate, assume that the normal distribution class is found suitable for expressing uncertainty about θ. The parameter ϕ then represents the parameters μ and σ. If the drilling operation were repeated a large number of times, or if there were a large number of operations similar to the one analysed, the parameters μ and σ could be observed as the mean and the standard deviation in this population, respectively. However, if the drilling of this specific well is considered as a unique operation, the parameters μ and σ cannot be observed. The parameters μ and σ are then fictional quantities interpreted as limiting quantities of a hypothetical population of identical situations generated by repeating the operation a large number of times.

Based on the information available in the planning phase, the uncertainty about ϕ is assessed. The prior density $f(\phi|H)$, reflecting the uncertainty about the value of the quantity ϕ, is assigned by use of experts, on the basis of the information H.

During the drilling operation, observations $x = (x_1, \ldots, x_n)$ are made and the assessment of ϕ should reflect the information gained from these observations. The probability density of ϕ when the nth observation is made is given by

$$f(\phi|x_n, \ldots, x_1, H) = cL(\phi|x_n) f(\phi|x_{n-1}, \ldots, x_1, H), \tag{15.4}$$

where

- $L(\phi|x_n)$ is the likelihood of ϕ given the observation x_n;

- $f(\phi|x_{n-1}, \ldots, x_1, H)$ is a probability density reflecting the value of ϕ given the information H and the observations $x_{n-1}, \ldots, x_1$.

The likelihood $L(\phi|x_n)$ is determined by assigning the probability density $f(x_n|\phi)$ for different values of ϕ. The probability density $f(x_n|\phi)$ for different values of ϕ. The probability densities $f(x_n|\phi)$, reflecting the uncertainty about the outcome of the nth observation given specific values of ϕ, are assigned by use of experts in the planning phase of the well.

However, the quantity of interest is not the parameter ϕ, but the pore pressure θ at depth d. Hence, after each assessment of ϕ, the probability density reflecting the uncertainty about θ should be calculated. When the nth observation is made, the probability density for θ is given by

$$f(\theta|x_n, \ldots, x_1, H) = \int_{\phi} f(\theta|\phi) f(\phi|x_n, \ldots, x_1, H) d\phi \qquad (15.5)$$

To summarize, the first step is to identify a probability model, $f(\theta|\phi)$. Then uncertainty about the quantity ϕ is assessed on the basis of the information H, leading to $f(\phi|\mathrm{H})$. When observations x are made, the assessment of ϕ is updated to obtain $f(\phi|x_i, \ldots, x_1, H)$, using Bayes' theorem. After each update, the uncertainty about θ is assessed and we obtain $f(\theta|x_i, \ldots, x_1, H)$. The difference between this alternative updating procedure and the procedure presented first is that the alternative updating procedure focuses on updating assessments of the parameter ϕ, instead of updating the assessment of θ directly.

The introduction of the probability model with parameter ϕ and the assessment of ϕ instead of θ, is motivated by the desire to simplify the probability assignments. However, if the parameter ϕ cannot be given a physical interpretation, the contrary is achieved. And for most wells this is the case; the drilling operation is regarded as a unique operation and the parameter ϕ cannot be given a physical interpretation. Even though a number of wells are drilled or planned to be drilled in the same area, there will always be details distinguishing one well from another. A population of similar operations can be defined, but we have to extend the meaning of 'similar' to a very broad class. The parameter ϕ is then of less relevance. In such cases, assessing the quantity θ directly is easier than introducing the probability model with the parameter ϕ.

Introducing the parameter ϕ also complicates the understanding and the assignment of the likelihood function. In the first part of the analysis, the likelihood function was closely related to the reliability of the measurement equipment used to perform the observation. The likelihood of x_i for a given value of ϕ is much more difficult to understand, let alone specify.

Remarks

The updating procedure presented in this section utilizes Bayes' theorem to update uncertainty assessments in light of new information, and this might give the impression that Bayes' theorem is the only way to account for new observations. However, this is not the case. For the problem presented above there is a need for a fast and automatic updating procedure, and Bayes' theorem is an appropriate method. In many cases, though, new information requires a rethinking of the whole information basis including the uncertainty assessments and the modelling, and Bayes' theorem is not appropriate (see the example below).

In the above analysis we recommend a direct assessment of the uncertainties about the quantity θ; we should not introduce a probability model $f(\theta|\phi)$ as a part of the updating procedure. The introduction of a probability model is motivated by the desire to simplify the assessments. However the contrary result is obtained when the parameters of the probability model are fictional quantities that cannot be given a physical interpretation. Fictional quantities can be difficult to understand and relate to, and this can complicate the assessment process substantially.

Assessing the number of events

Finally we consider a process plant example. As a part of the risk analysis of the plant, a separate study is to be carried out of the risk associated with the operation of the control room that is situated in a compressor module. Two persons operate the control room. The purpose of the study is to assess the risk to the operators as a result of possible fires and explosions in the module and to evaluate the effect of implementing risk-reducing measures. Based on the study, a decision will be made on whether to move the control room out of the module or to implement some risk-reducing measures. The risk is currently considered to be too high, but the management is not sure what is the overall best arrangement taking into account both safety and economy. The management decides to conduct a risk analysis to support the decision-making. To simplify, suppose the analysis is based on one event tree as shown in Figure 15.3. The tree models the possible occurrence of gas leakages in the process plant during a period of time, say 1 year. The number of gas leakages, referred to as the 'initiating events', is denoted by X. If an initiating event I occurs, it leads to N fatalities, where $N = 2$ if the events A and B occur, $N = 1$ if the events A and not B occur, and $N = 0$ if the event A does not occur. We may think of the event A as representing ignition

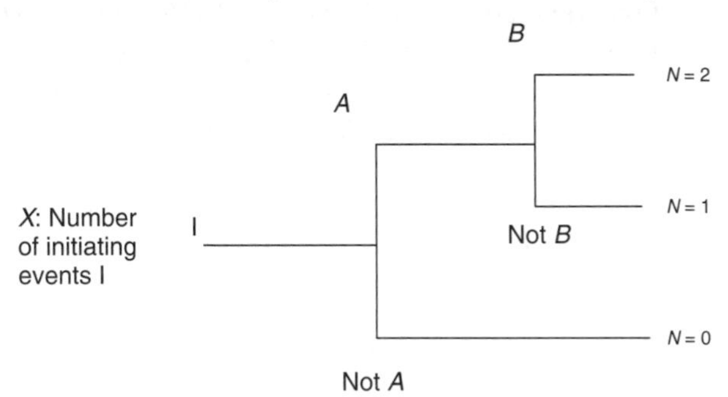

Figure 15.3 An event tree example (Aven 2003).

of the gas and B as explosion. The total number of fatalities is denoted by Y. To simplify the analysis we assume that the probability of two or more leakages during the period considered is negligible. Hence $N = Y$.

As an alternative to the above interpretation of the event tree, you may think of the initiating event as the failure in a production system, and A and B as failures of the barriers trying to reduce the consequences of this failure. Then the worst scenario is $N = 2$, the second worst is $N = 1$, and $N = 0$ is supposed to have no negative consequences.

Comprehensive textbook Bayesian approach

We begin by specifying a probability model for the problem, consisting of probability distributions for random variables (quantities) and parameters. First consider the number of initiating events X. A typical choice is to use a Poisson distribution $p(x|\lambda)$ with parameter λ, where

$$p(x|\lambda) = (\lambda^x/x!)\exp(-\lambda), \quad x = 0, 1, 2, \ldots.$$

Further, a prior distribution $f(\lambda|H)$ is specified, using for instance a gamma distribution. Here H denotes the background information that the assessment is based on. This information might come from various sources, for example more or less relevant historical data from similar situations, expert judgements, and so on. The gamma distribution is a mathematically convenient choice for prior distribution in this case as it a so-called conjugate distribution. A prior distribution is called conjugate if it leads to a posterior distribution in the same distribution class (in this case gamma). See, for example, Singpurwalla (2006) for more details.

Next we introduce $\theta_1 = \mathrm{P}(A|I)$ and $\theta_2 = \mathrm{P}(B|A)$. Then

$$\mathrm{P}(N = 2|I, \theta_1, \theta_2, H) = \theta_1\theta_2,$$

$$\mathrm{P}(N = 1|I, \theta_1, \theta_2, H) = \theta_1(1 - \theta_2),$$

$$\mathrm{P}(N = 0|I, \theta_1, \theta_2, H) = 1 - \theta_1,$$

and prior distributions (expressed as densities) $f(\theta_1|H)$ and $f(\theta_2|H)$ are specified, using for instance beta distributions. Note that more generally a joint prior distribution $f(\lambda, \theta_1, \theta_2, |H)$ of all parameters could be specified, but it is most common to use independent prior distributions, that is, $f(\lambda, \theta_1, \theta_2, |H) = f(\lambda|H)f(\theta_1|H)f(\theta_2|H)$.

If additional information in the form of observations $\mathrm{w} = (x_1, n_1, \ldots, x_n, n_k)$ of X and N (not previously included in H) is or becomes available, the likelihood function is given by

$$L(\lambda, \theta_1, \theta_2, |w, H) = \prod_i p(x_i|\lambda)\ \mathrm{P}(N = n_i|x_i, \theta_1, \theta_2),$$

where $\mathrm{P}(N = 2|x_i, \theta_1, \theta_2) = x_i\theta_1\theta_2$ and $\mathrm{P}(N = 1|x_i, \theta_1, \theta_2) = x_i\theta_1(1 - \theta_2)$.

The prior distributions can now be updated to posterior distributions by using Bayes' theorem. For instance,

$$f(\lambda|w, H) = cL(\lambda, \theta_1, \theta_2|w, H)f(\lambda|H),$$

where the constant of proportionality ensures that the posterior distribution is a proper distribution, that is, that it integrates to 1.

Some observations on the interpretation of the parameters are in order. A parameter is the limit of a function of so-called exchangeable observations. This interpretation follows from the so-called representation theorem (Bernardo and Smith, 1994). A sequence of observations is exchangeable if the joint distribution of any finite subsequence of observations is invariant under permutations of the order of the observations. Thus, in our example, λ is interpreted as the limit of the average of observations of the number of leakages, while θ_1 and θ_2 are interpreted as limiting frequencies of 0–1 events.

The parameters are fictional, based on thought experiments. The situations considered are not repeated a number of times under similar conditions – the situations are unique. Consider, for instance, the parameter λ in the example. This parameter is interpreted as the true expected number of leakages or as the limit of the average of the number of leakages, that is, λ is interpreted as a parameter of a thought-constructed hypothetical population. The same applies to θ_1 and θ_2 which are interpreted as probabilities or limiting frequencies of 0–1 events. There is no way that we can accurately measure these quantities, and introducing them produces uncertainties in the sense that the analysis has generated unknown quantities which do not exist in the real world. Obviously, care has to be taken when introducing such parameters if they cannot be given a meaningful interpretation.

An alternative Bayesian approach

An analysis along the above lines is seldom done in practice. Why do we introduce the probabilistic models and the fictional parameters λ, θ_1 and θ_2? The key quantities of interest are X and N, representing the number of leakages and the number of fatalities, respectively. These quantities should be predicted and associated uncertainties assessed.

The main focus is on the number of fatalities. To predict this number and to assess uncertainties, we develop a model, which is the event tree model shown in Figure 15.3. The aim of the modelling is to gain insight into the uncertainties and provide a tool for assessing the uncertainties about N. Given the model, the remaining uncertainties are related to the observable quantities X, A and B. The next step is then to assess the uncertainties of these quantities. Let us look at X first. We would like to predict X and assess the uncertainties. How should we do this? Suppose data from situations 'similar' to the one analysed are available, and let us assume for the sake of simplicity that these are of the form $x_1, x_2, \ldots, x_r$, where x_i is the number of initiating events during year i. These data are considered relevant for the situation being studied.

The data allow a prediction of X simply by using the mean x^* of the observations $x_1, x_2, \ldots, x_r$. But what about uncertainties? How should we express uncertainty related to X and the prediction of X? Suppose that the observations $x_1, x_2, \ldots, x_r$ are 1, 1, 2, 0, 1, so that $r = 5$ and the observed mean is equal to 1. In this case we have rather strong background information, and we suggest using the Poisson distribution with mean 1 as our uncertainty distribution of X. How can this uncertainty distribution be 'justified'? Well, if this distribution reflects our uncertainty about X, it is justified, and there is nothing more to say. This is a knowledge-based probability distribution and there is no need for further justification. But is a Poisson distribution with mean 1 'reasonable', given the background information? We note that this distribution has a variance not greater than 1. By using this distribution, 99% of the mass is on values less than 4.

Adopting the prevailing Bayesian thinking, as outlined above, using the Poisson distribution with mean 1, means that we have no uncertainty about the parameter λ, which is interpreted as the long-run average number of failures when considering an infinite number of exchangeable random quantities, representing systems similar to the one being analysed. According to the Bayesian theory, ignoring the uncertainty about λ gives misleading over-precise inference statements about X (Bernardo and Smith, 1994, p. 483). This reasoning is, of course, valid if we work within the standard Bayesian setting, considering an infinite number of exchangeable random quantities. In our case, however, we just have one X, so what do we gain by making a reference to limiting quantities of a sequence of similar hypothetical Xs? The point is that given the observations $x_1, x_2, \ldots, x_5$, the choice of the Poisson distribution with mean 1 is, in fact, reasonable. Consider the following argumentation. Suppose that we divide the year $[0, T]$ into time periods of length T/k, where k is, for example, 365 or 1000. Then we may ignore the possibility of having two events occurring in one time period, and we assign an event probability of $1/k$ for the first time period, as we predict one event in the whole interval $[0, T]$. Suppose that we have observations related to $i - 1$ time periods. Then for the next time period we should take these observations into account. Using independence means ignoring available information. A natural way of balancing the prior information and the observations is to assign an event probability of $(d_i + 1 \cdot r)/((i - 1) + rk)$, where d_i is equal to the total number of events that occurred in $[0, T(i - 1)/k]$, that is, we assign a probability which is equal to the total number of events occurring per unit of time. It turns out that this assignment process gives an approximate Poisson distribution for X. This can be shown, for example, by using Monte Carlo simulation. The Poisson distribution is justified as long as the background information dominates the uncertainty assessment of the number of events occurring in a time period. Thus, from a practical point of view, there is no problem in using the Poisson distribution with mean 1. The above reasoning provides a 'justification' for the Poisson distribution, even with not more than one or two years of observations.

Note that for the direct assignment procedure using the k time periods, the observations $x_1, x_2, \ldots, x_5$ are considered a part of the background information, meaning that this procedure does not involve any modelling of these data. In contrast, the more standard Bayesian approach requires that we model $x_1, x_2, \ldots, x_5$ as observations coming from a Poisson distribution, given the mean λ.

We conclude that a Poisson distribution with mean 1 can be used to describe the analyst's uncertainty with respect to X in this case. The background information is sufficiently strong.

The above analysis is a tool for predicting X and assessing associated uncertainties. When we have little data available, modelling is required to gain insight into the uncertainty related to X and hopefully reduce the uncertainty. The modelling also makes it possible to see the effects of changes in the system and to identify risk contributors.

We now turn to how to assess uncertainties for A and B. For these events we just need to specify two knowledge-based probabilities, $P(A|I)$ expressing our uncertainty related to occurrence of ignition, and $P(B|A)$ expressing our uncertainty regarding an explosion given an ignition. The basis for the probability assignments would be 'hard' data and expert opinions. These probabilities are not 'true underlying probabilities' or limiting frequencies of 0–1 events, they just represent our knowledge-based uncertainties regarding the observable events A and B, expressed as probabilities. This is different from the above textbook Bayesian analysis where prior distributions expressing the uncertainties regarding the 'true values' of θ_1 and θ_2 are usually specified. Why introduce such hypothetical limiting quantities (chances) and associated prior distributions, when we can easily assess our uncertainties regarding what would happen by the single numbers $P(A|I)$ and $P(B|A)$?

What remains now is to use probability calculus to calculate the uncertainty distribution for the number of fatalities N. So the end product of the analysis is simply the uncertainty distribution expressing our uncertainty regarding the future value of N. There are no further 'uncertainties about uncertainties'. The uncertainty distribution regarding the 'top-level' quantity, here N, is calculated by first focusing on the observable quantities on a more 'detailed level', in this case X, and A and B, establishing uncertainty distributions for these, and then using probability calculus to propagate this into an uncertainty distribution for the top-level quantity N.

Remarks

The key principles followed in the alternative approach are as follows. The focus is on observable quantities, that is, quantities expressing states of the 'world' or nature, that are unknown at the time of the analysis but will (or could) become known in the future (Aven, 2003). These quantities are predicted in the risk analysis and probability is used as a measure of uncertainty related to the true values of these quantities. The emphasis on these principles gives a framework which is easy to understand and use in a decision-making context. As for the practical calculations, compared to the traditional approach there is no need for 'prior distributions' on quantities like θ_1, θ_2 and λ.

If an infinite (or large) population of similar situations can be defined, then the parameters represent a state of nature – they are observable quantities – and we can speak about our uncertainty about these quantities. The question is whether such a population of similar situations should be introduced. In our view, as a general rule, the analyst should avoid introducing fictitious populations. In the example considered above, it is not obvious how to define an infinite or large population of similar situations, and without a precise understanding of the population, the uncertainty assessments become difficult to perform and it introduces an element of arbitrariness. What is a fictitious population and what is a real population is a matter for the analyst to decide, but the essential point we are making here is that the analyst should think first before introducing such a population. The full Bayesian set-up, covering the introduction of a parametric distribution class, specification of a prior distribution for the parameters, Bayesian updating to establish the posterior distribution, and the calculation of the predictive distributions, is a useful tool for coherent analysis of statistical data, but should not be used when not appropriate or needed.

To apply Bayesian updating some sort of model stability is required as discussed above: populations of similar units need to be constructed (formally an infinite set of exchangeable random variables). But such stability is often not fulfilled. See Bergman (2009) for further reflections on this issue.

Summary

Bayesian updating is a powerful tool for systematic incorporation of prior knowledge and observational data. The approach is based on the definition of probabilistic models which include unknown parameters. When to introduce such models is a judgement to be made by the analyst and needs to be seen in relation to the purpose of the analysis. For problems where continuous updating is required, such modelling ensures a systematic analysis of the uncertainties, but such models should be used with care when the parameters cannot be given meaningful interpretations. Quantitative uncertainty assessments of vaguely defined quantities make no sense. The focus should be on the observable quantities and the prediction and assessment of these. Probabilistic models could be introduced but only if they can be justified.

References

Aven, T. (2003) *Foundations of Risk Analysis*, John Wiley & Sons, Ltd, Chichester.

Bergman, B. (2009) Conceptualistic pragmatism: a framework for Bayesian analysis? *IIE Transactions*, **41**, 86–93.

Bernardo, J. and Smith, A. (1994) *Bayesian Theory*, John Wiley & Sons, Ltd, Chichester.

Sandøy, M. and Aven, T. (2006) Real time updating of risk assessments during a drilling operation. *International Journal of Reliability, Quality and Safety Engineering*, **13**, 85–95.

Singpurwalla, N. (2006) *Reliability and Risk. A Bayesian Perspective*, John Wiley & Sons, Ltd, Chichester.

Further reading

Aven, T. and Kvaløy, J.T. (2002) Implementing the Bayesian paradigm in risk analysis. *Reliability Engineering and System Safety*, **78**, 195–201.

16

Sensitivity analysis is a type of uncertainty analysis

A sensitivity analysis in a risk analysis context is a study of how sensitive a calculated risk index is with respect to changes in conditions and assumptions made. Figure 16.1 shows an example of a sensitivity analysis of the calculated expected profit of a project. In the analysis the oil price was initially set to $50 per barrel, but the oil price is subject to large uncertainties and it obviously provides insights to show how the results depend on the oil price. The actual oil price could deviate strongly from the initial assumption of $50 per barrel.

Many analysts refer to such a sensitivity analysis as an uncertainty analysis (with respect to the oil price). However, the sensitivity analysis does not include any assessment of the uncertainties about the oil price, which is considered necessary for the analysis to be referred to as an 'uncertainty analysis'. The sensitivity analysis provides a basis for an uncertainty analysis with respect to the oil price. If a probability distribution is assigned for the oil price, we can use the sensitivity analysis to produce an unconditional expected profit. Alternatively, the uncertainty assessment could be a qualitative assessment of the oil price and the results are reported together with Figure 16.1.

To give a more detailed example of a sensitivity analysis, we consider the development of a model for analysing the reliability and lifetime of a technical system which comprises two components (labelled 1 and 2) in parallel (see Figure 13.1). The reliability h of the system is given by

$$h(p) = 1 - (1 - p_1)(1 - p_2),$$

where p_i is the reliability of component i, $i = 1, 2$. We interpret h and p_i as the proportion of functioning systems and components when considering an infinite number of similar systems and components. Suppose that the reliability analysis produces values $p_1 = 0.60$ and $p_2 = 0.90$, which gives a system reliability

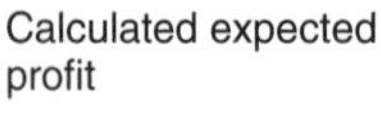

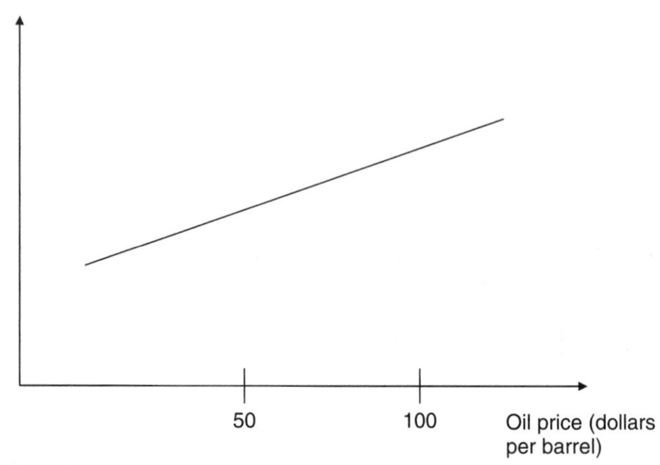

Figure 16.1 Example of a sensitivity analysis.

$h = 0.96$, assuming independent components. To see how sensitive the system reliability is with respect to variations in the component reliabilities, a sensitivity analysis is carried out. Figure 16.2 shows the results. The upper line shows the system reliability h when the component reliability of component 1 varies from 0 to 1 and component 2 has reliability 0.90. The lower line shows the system reliability h when the component reliability of component 2 varies from 0 to 1 and component 1 has reliability 0.60. We see that that relative improvement in system reliability is largest for component 2. Hence, a small improvement of component 2 gives the largest effect on system reliability. The relative increase is given by the slope of the lines, which equals 0.1 and 0.4, respectively.

This information can be used in two ways:

1. It shows that the assignment of $p_2 = 0.90$ is more critical than the assignment of $p_1 = 0.60$ in the sense that deviations from these values would give strongest effect for component 2.

2. It shows that a small improvement of component 2 would have the strongest effect on the system reliability.

In this manner the sensitivity analysis is being used as a tool for ranking the components with respect to criticality and importance. A number of such criticality/importance measures have been presented in the literature. See, for example, van der Borst and Schoonakker (2001), Borgonovo *et al.* (2003) and Aven and Jensen (1999). In the above analysis the focus is the partial derivative of the reliability (object) function, that is, the relative change in $h(p)$ relative to a change in p_i:$\partial h/\partial p_i$. The measure is referred to as the Birnbaum importance measure.

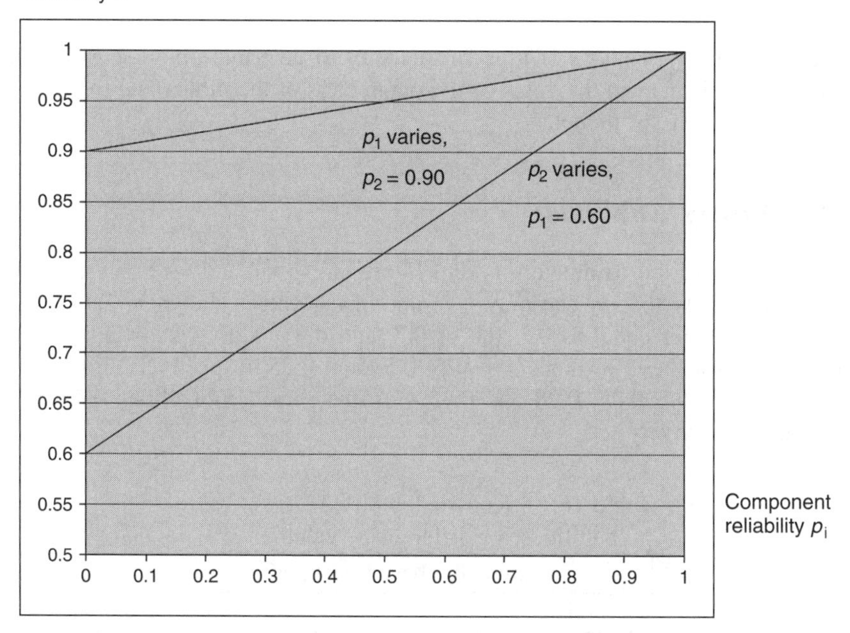

Figure 16.2 Sensitivity analysis with respect to the component reliabilities for the reliability function $h(p) = 1 - (1 - p_1)(1 - p_2)$.

Another common importance measure generated by the sensitivity analysis is the improvement potential or the achievement worth, defined by the change in the object function when p_i is set to its maximum value, 1. In our case this leads to the same ranking of the components as both have the potential to make the system perfectly reliable. However, in most situations this would not be the case and the measure provides useful information about where the largest improvements can be achieved.

Let us go one step further and model the component reliability p_i. Suppose that an exponential distribution $F_i(t) = 1 - \exp\{-\lambda_i t\}$ is used, where λ_i is the failure rate, $i = 1, 2$. The interpretation of the model is as follows: $F_i(t)$ represents the fraction of lifetimes less than or equal to t when considering an infinite (very large) population of similar units of component i, and $1/\lambda_i$ is the mean lifetime in this population.

In this model $p_i = \exp\{-\lambda_i t\}$, where t is a fixed point in time. If T_i denotes the lifetime of component i, we have $\mathrm{P}(T_i > t) = 1 - F_i(t) = \exp\{-\lambda_i t\}$. At the system level we introduce the lifetime T, and we can write

$$P(T > t) = 1 - (1 - \exp\{-\lambda_1 t\})(1 - \exp\{-\lambda_2 t\}).$$

Sensitivity analyses can now be conducted with respect to the failure rates λ_i and with respect to the choice of the exponential distribution. An alternative model is the Weibull distribution with parameters α and β, and by varying these parameters we may study the sensitivity of the system reliability with respect to the model choice.

The sensitivity analysis is used in this way to determine how sensitive the results are to changes in the value of the parameters of the model and to changes in the structure of the model.

Uncertainty analysis

The true value of the parameters p_i and λ_i are unknown, and we assume for the sake of simplicity that p_i can only take one of the values shown in Table 16.1. Hence p_1 is either equal to 0.5, 0.6 or 0.7, and the analyst's knowledge-based probabilities for these p values are 0.25, 0.50 and 0.25, respectively. For p_2 the possible values are 0.85, 0.90 and 0.95 with the same probabilities 0.25, 0.50 and 0.25, respectively.

Table 16.1 Knowledge-based probabilities for different p_i values.

	Knowledge-based probabilities
p_1 values	
0.5	0.25
0.6	0.50
0.7	0.25
p_2 values	
0.85	0.25
0.90	0.50
0.95	0.25

Using these knowledge-based probabilities, we can calculate a probability distribution for the system reliability; see Table 16.2 and Figure 16.3. Independence in the assessment of p_1 and p_2 is used, for example $P(p_1 = 0.7 | p_2 = 0.95) = P(p_1 = 0.7)$.

We see from Table 16.2 and Figure 16.3 that a probability of 0.25 is computed for $h = 0.96$. The value 0.96 is the most likely outcome and it is also the expected value of h with respect to the uncertainty distributions of the p_i. This can be shown by computing the expected value directly from the data of Table 16.2, or more generally by noting that

$$\begin{aligned} E[h(p)] &= E[1 - (1 - p_1)(1 - p_2)] = 1 - (1 - Ep_1)(1 - Ep_2) \\ &= 1 - (1 - 0.6)(1 - 0.90) = 0.96. \end{aligned}$$

Table 16.2 Knowledge-based probabilities for the system reliability based on the input from Table 16.1.

h values	Knowledge-based probabilities
0.925	0.0625
0.940	0.125
0.950	0.125
0.955	0.0625
0.960	0.250
0.970	0.125
0.975	0.0625
0.980	0.125
0.985	0.0625

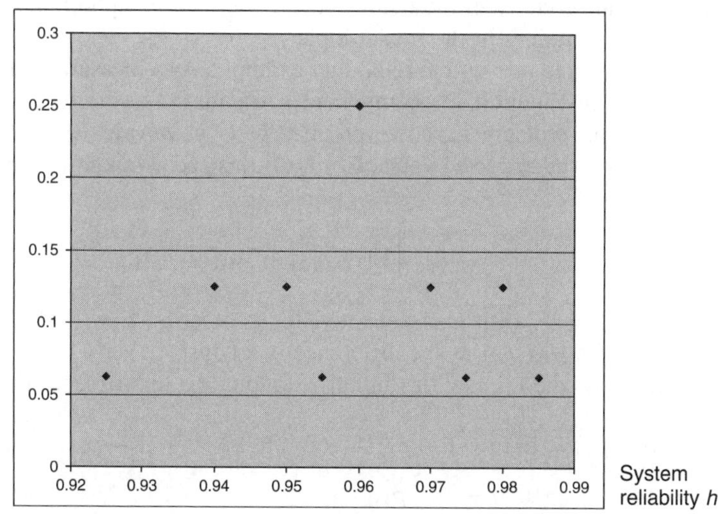

Figure 16.3 Probability distribution of h based on Table 16.2.

The true system reliability could, however, deviate from the expected value. Based on the distributions of p_1 and p_2, we see that the distribution of h has all its mass in the interval [0.925, 0.985]. The variance (standard deviation) is a measure of the spread of the distribution of h. Based on the data of Table 16.2 we compute

$$\text{Var}[h] = 0.000256 \text{ and } \text{SD}[h] = 0.0160.$$

We have performed an uncertainty analysis of the system reliability. Formally, an uncertainty analysis refers to the determination of the uncertainty in analysis results (here h) that derives from uncertainty in analysis inputs (here uncertainty about p_1 and p_2); see Helton *et al.* (2006). Using the notation introduced in Chapter 13, we may formulate the problem as follows. To analyse a quantity Z, we introduce a model $G(X)$, which depends on the input quantities X, and set $Z = G(X)$. An uncertainty analysis of Z means an assessment of the uncertainties about X and an uncertainty propagation through G to produce an assessment of the uncertainties about Z. Knowledge-based probabilities are used to express these uncertainties. In such a context it is also common to perform sensitivity analyses as will be discussed next.

Sensitivity analysis in the context of an uncertainty analysis

A sensitivity analysis in this context refers to the determination of the contributions of individual uncertainty analysis inputs to the analysis results (Helton *et al.*, 2006), or, as formulated by Saltelli (2002): sensitivity analysis is the study of how the uncertainty in the output of a model can be apportioned to different sources of uncertainty in the model input.

Let us return to our system reliability example. As a measure of the sensitivity of the input distribution for component i we may use the variance, $v_i(p_i) = \mathrm{Var}[h|p_i]$. The problem is, however, that p_i is unknown. It would thus seem sensible to use the expected value, $\mathrm{E}[v_i(p_i)]$, instead. We know from probability theory that

$$\mathrm{Var}[h] = \mathrm{E}[v_i(p_i)] + \mathrm{Var}[u_i(p_i)], \tag{16.1}$$

where $u_i(p_i) = \mathrm{E}[h|p_i]$. Consequently, for comparing $\mathrm{E}[v_i(p_i)]$ we may alternatively use $\mathrm{Var}[u_i(p_i)]$ as the difference is a constant $\mathrm{Var}[h]$.

To compute these quantities in our example, we first observe that

$$\begin{aligned}
\mathrm{E}[h|p_1] &= \mathrm{E}[1-(1-p_1)(1-p_2)|p_1] = 1-(1-p_1)(1-\mathrm{E}p_2) \\
&= 0.90 + 0.10\,p_1, \\
\mathrm{E}[h|p_2] &= \mathrm{E}[1-(1-p_1)(1-p_2)|p_2] = 1-(1-\mathrm{E}p_1)(1-p_2) \\
&= 0.60 + 0.40\,p_2.
\end{aligned}$$

From these expressions we obtain

$$\begin{aligned}
\mathrm{Var}[u_1(p_1)] &= \mathrm{Var}[0.90+0.10p_1] = 0.1^2\mathrm{Var}[p_1] = 0.010 \cdot 0.005 = 0.00005, \\
\mathrm{Var}[u_2(p_2)] &= \mathrm{Var}[0.60+0.40p_2] = 0.4^2\mathrm{Var}[p_2] = 0.16 \cdot 0.00125 = 0.00020.
\end{aligned}$$

Thus, the uncertainty distribution of component 1 has a stronger influence on the variation of the output result than the uncertainty distribution of component

2. Remember formula (16.1), which says that $\mathrm{Var}[u_i(p_i)]$ is equal to a constant minus $\mathrm{E}[v_i(p_i)]$. This means that if we are able to reduce the uncertainties of p_1 this would have a larger effect on the uncertainties of h than reducing the uncertainties of p_2.

An alternative sensitivity measure (importance measure) is the correlation coefficient ρ between p_i and h, defined by

$$\rho(p_i, h) = \frac{\mathrm{Cov}(p_i, h)}{\mathrm{SD}[p_i]\ \mathrm{SD}[h]}.$$

The correlation coefficient is a normalized version of the covariance. It takes values in $[-1, 1]$. Here $+1$ indicates perfect positive correlation and -1 perfect negative correlation. The index expresses the strength of the relationship between p_i and h. The numerical analysis gives

$$\rho(p_1, h) = \frac{\mathrm{Cov}(p_1, h)}{\mathrm{SD}[p_1]\ \mathrm{SD}[h]} = \frac{0.0005}{0.071 \cdot 0.016} = 0.44,$$

$$\rho(p_2, h) = \frac{\mathrm{Cov}(p_2, h)}{\mathrm{SD}[p_2]\ \mathrm{SD}[h]} = \frac{0.0004}{0.035 \cdot 0.016} = 0.71,$$

noting that

$$\begin{aligned}
\mathrm{Cov}(p_1, h) &= \mathrm{E}[p_1 h] - \mathrm{E}[p_1]\ \mathrm{E}[h] = \mathrm{E}[p_1^2(1 - p_2) + p_1 p_2] - \mathrm{E}p_1\ \mathrm{E}h \\
&= 0.365 \cdot 0.1 + 0.6 \cdot 0.9 - 0.6 \cdot 0.96 = 0.0005, \\
\mathrm{Cov}(p_2, h) &= \mathrm{E}[p_2 h] - \mathrm{E}[p_2]\ \mathrm{E}[h] = \mathrm{E}[p_2^2(1 - p_1) + p_1 p_2] - \mathrm{E}p_2\ \mathrm{E}h \\
&= 0.811 \cdot 0.4 + 0.6 \cdot 0.9 - 0.9 \cdot 0.96 = 0.0004, \\
\mathrm{Var}[p_1] &= \mathrm{E}[p_1^2] - (\mathrm{E}p_1)^2 = 0.365 - 0.360 = 0.005, \\
\mathrm{Var}[p_2] &= \mathrm{E}[p_2^2] - (\mathrm{E}p_2)^2 = 0.81125 - 0.8100 = 0.00125.
\end{aligned}$$

Thus the calculations show that the correlation coefficient is largest for component 2. This was expected as the component with the highest reliability dominates the reliability of the system.

A common approach for showing the sensitivities of the component input to the output is to use scatterplots, as in Figures 16.4 and 16.5. To produce these plots we have used revised distributions for p_1 and p_2, as shown in Table 16.3.

Within the intervals, for example 0.45–0.55, a uniform distribution is assumed, meaning that the knowledge-based probability that p_1 is less than (say) 0.47 equals 2/10 of 25%, or 5%. By drawing n random numbers on an Excel spreadsheet, we produce n values for p_1 and p_2 and h, as shown in Figures 16.4 and 16.5. We see from the plots that p_2 and h are much more dependent than p_1 and h, as was expected from the above analysis.

Tornado charts for different methods of sensitivity analysis are also in common use. One is based on a correlation measure, for example the correlation

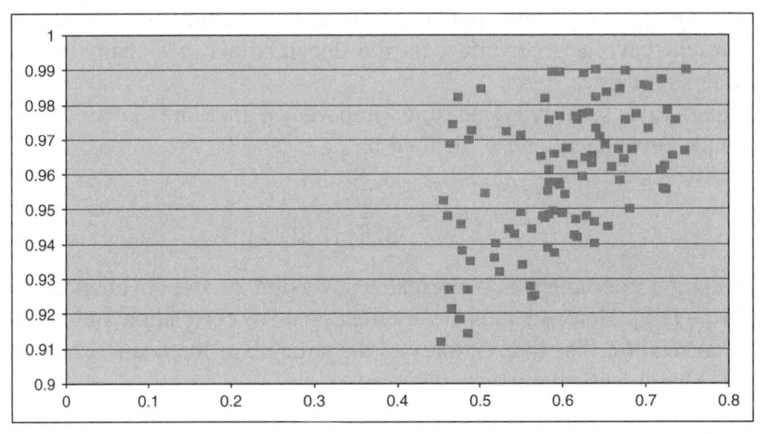

Figure 16.4 Scatterplot for p_1 and h.

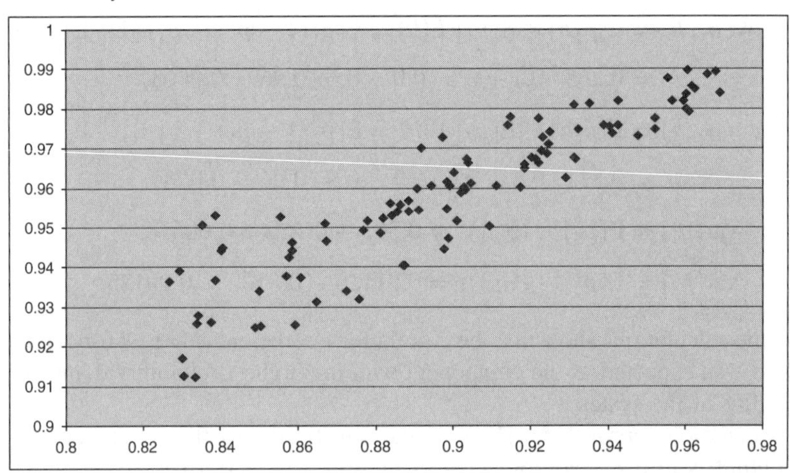

Figure 16.5 Scatterplot for p_2 and h.

coefficient ρ computed above. An illustrative example is shown in Figure 16.6 for a petroleum project (based on Vose 2008, p. 83). Another chart is based on computing the expected output for each input parameter fixed to a low value (say, its 5% quantile) and a high value (say, its 95% quantile); see Figure 16.7.

Table 16.3 Revised knowledge-based probabilities for different p_i values.

	Knowledge-based probabilities
p_1 values	
0.45–0.55	0.25
0.55–0.65	0.50
0.65–0.75	0.25
p_2 values	
0.825–0.875	0.25
0.875–0.925	0.50
0.925–0.975	0.25

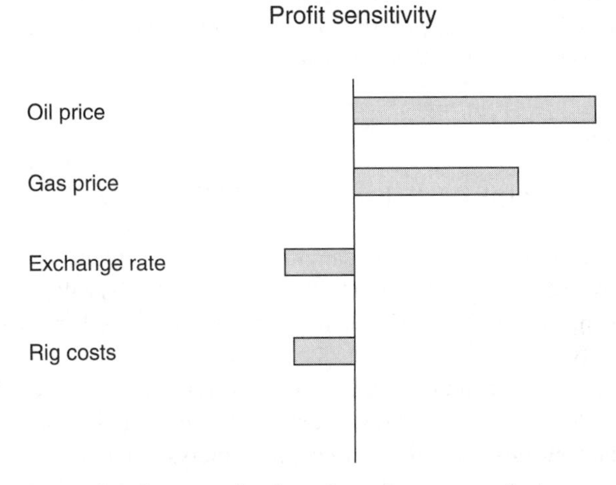

Figure 16.6 Example of a tornado chart based on a correlation measure between various parameters and the profit of a petroleum project. (based on Vose, 2008, p. 83).

For our reliability example, such an analysis gives the following intervals for the system reliability:

- component 1: [0.947, 0.973]
- component 2: [0.934, 0.986].

To establish these intervals, we first define the 5 and 95% quantiles:

- component 1: 0.47, 0.73
- component 2: 0.835, 0.965.

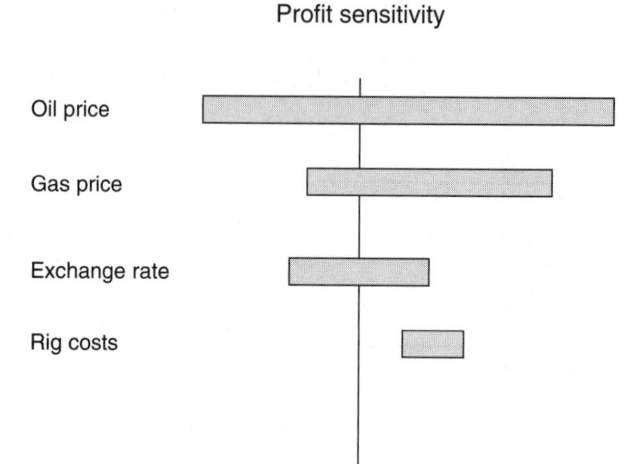

Figure 16.7 Example of a tornado chart expressing the expected profit variation using the 5% and 95% quantiles of the uncertainty distribution of various parameters. (based on Vose, 2008, p. 83).

Then we obtain

$$E[h|p_1 = 0.47] = 0.47 + 0.53Ep_2 = 0.947.$$

The other results follow by similar calculations. These intervals are depicted in Figure 16.8. We see that component 2 has the widest intervals, which indicates that variation in the reliability for this component has the largest effect on the system reliability.

A number of other techniques exist for sensitivity analysis; see, for example, Helton *et al.* (2006) and Vose (2008). Many of these are based on sampling from probability distributions as in the scatter plots analysis above.

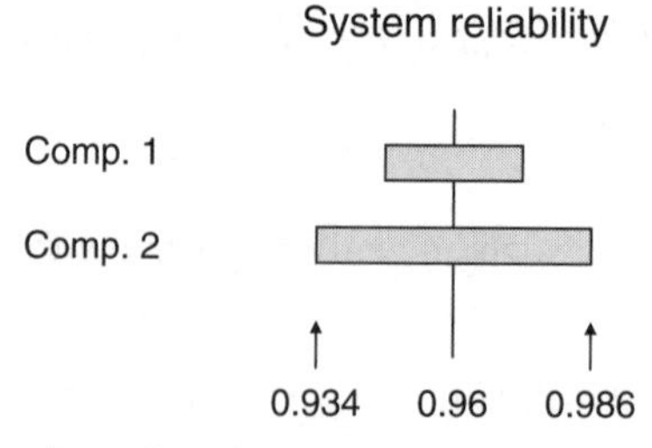

Figure 16.8 Tornado chart for the system reliability example, expressing the expected system reliability using the 5% and 95% quantiles of the uncertainty distribution of p_i.

Reflection

We consider a parallel system of two components, as shown in Figure 13.1. The state Z of the system is equal to 1 if the system is functioning and 0 otherwise. The system is functioning if at least one of the components is functioning. We adopt knowledge-based probabilities and, using independence, we obtain

$$P(Z = 1|K) = 1 - (1 - p_1)(1 - p_2),$$

where $p_i = P(X_i = 1|K)$ and K is the background knowledge. The analysis is based on the model G which describes the parallel system:

$$G(X) = 1 - (1 - X_1)(1 - X_2),$$

where X_i is the state of component i, $i = 1, 2$, defined as 1 if component i is functioning and 0 otherwise. Further modelling is not introduced.

A component importance measure is to be defined. How would you define this measure?

The answer depends on what you intend this measure to reflect. Obviously, different situations and objectives call for different measures. If we are concerned about small changes in a component's reliability we could use the Birnbaum measure, whereas if we are more interested in identifying the maximum potential of improving a component's reliability, we are led to the improvement potential. The uncertainties are reflected in the p_i and there is no point in introducing a second level of uncertainties as in the analysis above where we introduced chances, interpreted as proportions in infinite populations of similar units. It is obviously essential to be precise on what the quantities of interest are, Z or the chance h.

Summary

Uncertainty analysis refers to the determination of the uncertainty in analysis results that derives from uncertainty in analysis inputs. Sensitivity analysis in this context refers to the determination of the contributions of individual uncertainty analysis inputs to the analysis results. An example of such a sensitivity measure is the correlation coefficient. The context for this analysis is unknown quantities X and output quantities $Z = G(X)$, and the uncertainties of these quantities are expressed through knowledge-based probabilities.

Sensitivity analysis is also used to show how the variation of a quantity x_i influences a probabilistic index for Z, for example EZ. In this case the uncertainties about x_i are not taken into account in the analysis. Thus, the analysis is not an uncertainty analysis.

References

Aven, T. and Jensen, U. (1999) *Stochastic Models in Reliability*, Springer-Verlag, New York.

Borgonovo, E., Apostolakis, G.E., Tarantola, S. and Saltelli, A. (2003) Comparison of global sensitivity analysis techniques and importance measures in PSA. *Reliability Engineering and System Safety*, **79**, 175–185.

Helton, J.C., Johnson, J.D., Sallaberry, C.J. and Storlie, C.B. (2006) Survey of sampling-based methods for uncertainty and sensitivity analysis. *Reliability Engineering and System Safety*, **91**, 1175–1209.

Saltelli, A. (2002) Sensitivity analysis for importance assessment. *Risk Analysis*, **22**, 579–590.

van der Borst, M. and Schoonakker, H. (2001) An overview of PSA importance measures. *Reliability Engineering and System Safety*, **72**, 241–245.

Vose, D. (2008) *Risk Analysis*, 3rd edn, Chapters 5 and 13, John Wiley & Sons, Ltd, Chichester.

Further reading

Aven, T. and Nøkland, T.E. (2010) On the use of uncertainty importance measures in reliability and risk analysis. *Reliability Engineering and System Safety*, **95**, 127–133.

17

The main objective of risk management is risk reduction

A common conception is that the main objective of risk management is to reduce risk by ensuring that adequate measures are taken to protect people, the environment and assets from the harmful consequences of the activities being undertaken. But such a perspective can be challenged. Consider the following examples.

Example. Exploration of space

What is the aim of risk management in the context of space exploration? To reduce the risk of loss of human lives? No, of course not. If that had been the main goal, no manned space missions should have been initiated as such missions would obviously involve high risks of loss of lives. The driving force for sending human beings in spacecraft is not safety and risk reduction but a genuine willingness to use resources to explore space and to expand into all possible niches. All 'successful civilizations' have explored their surrounding areas and faced risks. The achievements are often astonishing, like sending a man to the moon: nearly 50 years ago (25 May 1961), President John F. Kennedy presented a bold challenge before a joint session of Congress: to send a man to the moon by the end of the decade (CNN, 2001). 'No single space project in this period will be more impressive to mankind, or more important for the long-range exploration of space; and none will be so difficult or expensive to accomplish', Kennedy said.

Despite sceptics who thought it could not be accomplished, Kennedy's dream became a reality on 20 July 1969, when the Apollo 11 commander Neil Armstrong took 'one small step for a man, a giant leap for mankind', leaving a dusty trail of footprints on the moon.

Misconceptions of Risk T. Aven

The space programmes have not been without accidents and deaths, but most people would probably judge the success rate as extremely impressive. Although the astronauts' safety has always been a main concern, no one would expect that such a bold task could be realized without some accidents. Accidents are the price of the exploration. It is not destiny, but a consequence of the fact that the task is difficult and the risks are high.

Risk reduction enters the arena when a decision has been made to perform a given activity: sending a man to the moon. The issue is then how to realize the project and ensure sufficiently low risk for the astronauts. Different technological concepts may be considered in the process, and their suitability would be judged by reference to a number of attributes, including reliability and costs. One concept may have a large potential for risk reduction but could be rejected as it is too costly and causes many years of delay to the project.

The spacecraft is then built and the space mission is to be accomplished. A number of factors, for example related to the training and competence of the astronauts, would be critical for the successful completion of the mission. Risk reduction is a main objective of risk management at this stage of the project.

Example. Investment in securities

An investor considers two investment strategies:

A. Invest $\$x$ million in securities of type A which are characterized by high expected profit but also large uncertainties.

B. Invest the same amount in securities of type B which are characterized by considerably lower expected profit but also a considerably lower level of uncertainties.

If risk management were all about reducing risk, strategy B would be favourable, but of course, such a conclusion cannot be made in general. Depending on the investor's attitude to risk and uncertainties, strategy A could be preferred. Risk reduction may be desirable, but not without considering its value.

Example. Oil and gas exploration

Let us go back to the start of offshore petroleum activities on the Norwegian Continental Shelf in the North Sea in the late 1960s and the beginning of the 1970s (see Aven and Vinnem, 2007). If safety and the avoidance of fatal accidents had been the main goal, the activity should obviously not have been started. Many accidents have occurred in offshore oil and gas exploration and any assessment would have indicated a very high probability of many hundreds of fatalities in the coming years. Nonetheless, politicians gave their consent for the activity to proceed. The potential benefits to society were enormous. The economic incentives were so strong that the politicians 'simply had no choice'. Few people would probably say that the government made the wrong decision, even though several hundred have lost their lives in the industry. Norwegian society would probably have been in a completely different position than it is today, if the government had not been willing to accept the risks.

As for exploration in space, the risk management would be different when the task is to implement the activity and operate the installations. □

Based on these examples we distinguish between

(i) risk linked to the realization of the activity

(ii) risk linked to constraints such as concept selection and technology

(iii) risk linked to 'free variables' within the constraints of (i) and (ii). These free variables are often related to human and organizational factors.

Using this classification, we may respond to the question about risk reduction being an objective negatively in case (i), both negatively and positively for (ii) and positively for (iii).

Increased activity generally leads to higher risks, so it would be meaningless to say that risk reduction is the main objective in the case of (i). When it comes to (ii) and (iii), risk reduction is, however, an objective, but there will always be other concerns to be balanced. We will discuss this in more detail in the following. First we need to formulate a more suitable objective for risk management.

Basic risk management theory

Any organization exists to provide value for its stakeholders. A challenge for management is to determine how much uncertainty and risk to accept as it strives to grow stakeholder value. It is the task of risk management to strike the right balance between exploring opportunities on the one hand, and avoiding losses, accidents and disasters on the other (Aven, 2008). It is about balancing growth and return goals and related uncertainties, and efficiently and effectively deploying resources in pursuit of the entity's objectives (COSO, 2004). Risk management relates to all activities, conditions and events that can affect the organization, and its ability to reach its goals and vision.

Risk management encompasses the following (COSO, 2004):

- *Aligning risk appetite and strategy*. Management considers the entity's risk appetite in evaluating strategic alternatives, setting related objectives and developing mechanisms to manage related risks.
- *Enhancing risk response decisions*. Risk management provides the rigour to identify and select among alternative risk responses – risk avoidance, reduction, sharing and acceptance.
- *Reducing operational surprises and losses*. Entities gain enhanced capability to identify potential events and establish responses, reducing surprises and associated costs or losses.

- *Identifying and managing multiple and cross-organizational risks*. Every organization faces a myriad of risks affecting different parts of the organization, and risk management facilitates an effective response to the interrelated impacts, and integrated responses to multiple risks.

- *Seizing opportunities*. By considering a full range of potential events, management is positioned to identify and proactively realize opportunities.

- *Improving deployment of capital*. Obtaining robust risk information allows management to effectively assess overall capital needs and enhance capital allocation.

These capabilities inherent in risk management help management achieve the entity's performance and profitability targets and prevent loss of resources. Risk management helps ensure effective reporting and compliance with laws and regulations, and helps avoid damage to the entity's reputation and associated consequences. In sum, risk management helps an entity get to where it wants to go and avoid pitfalls and surprises along the way.

It is common to distinguish between management strategies for handling the risk agent (such as a chemical or a technology) and those needed for the risk-absorbing system (such as a building, an organism or ecosystem); cf. Table 7.1 (IRGC, 2005; Aven and Renn, 2009). With respect to risk-absorbing systems robustness and resilience are two main categories of strategies/principles in the case of large uncertainties. Robustness refers to the insensitivity of performance to deviations from normal conditions. Measures to improve robustness include inserting conservatisms or safety factors as an insurance against individual variation, introducing redundant and diverse safety devices to improve structures against multiple stress situations, reducing the susceptibility of the target organism (e.g. iodine tablets for radiation protection), establishing building codes and zoning laws to protect against natural hazards, as well as improving the organizational capability to initiate, enforce, monitor and revise management actions (high reliability, learning organizations).

With respect to risk-absorbing systems, an important objective is to make these systems resilient so they can withstand or even tolerate surprises. In contrast to robustness, where potential threats are known in advance and the absorbing system needs to be prepared to face these threats, resilience is a protective strategy against unknown or highly uncertain events. Instruments for resilience include the strengthening of the immune system, diversification of the means for approaching identical or similar ends, reduction of the overall catastrophic potential or vulnerability even in the absence of a concrete threat, design of systems with flexible response options, and the improvement of conditions for emergency management and system adaptation. Robustness and resilience are closely linked but they are not identical and require partially different types of actions and instruments.

Recently many standards and textbooks on risk management have been issued, and there is, to a large extent, agreement among them when it comes to these overall features; see, for example, AS/NZS (2004), COSO (2004), ISO (2008),

Jorion (2007) and Aven and Vinnem (2007). Integrated frameworks, often referred to as enterprise risk management and integrated risk management, are introduced to implement the risk management in organizations. These frameworks have an organization-wide perspective and cover the total risk picture, that is, all risk aspects of concern for the organization.

Risk analysis has an important role in risk management. The analyses include identification of hazards and threats (and opportunities), cause analyses, consequence analyses and risk description. The results of the analyses are then evaluated, that is, compared with possible criteria and reference levels defined. Together, the analyses and the evaluations are referred to as risk assessments. Risk assessment is followed by risk treatment, which is a process involving the development and implementation of measures to modify risk, including measures designed to avoid, reduce ('optimize'), transfer or retain risk. Risk transfer means sharing with another party the benefit or loss associated with a risk. It is typically effected through insurance.

Risk management is an integral part of good management practice and an essential element of good corporate governance. It is an iterative process consisting of steps that, when undertaken in sequence, enable continuous improvement in decision-making and facilitate continuous improvement in performance, within the limits discussed above related to the balance of different concerns.

Generally, the risk management process can be divided into four main steps (IRGC, 2005) as presented in Figure 17.1:

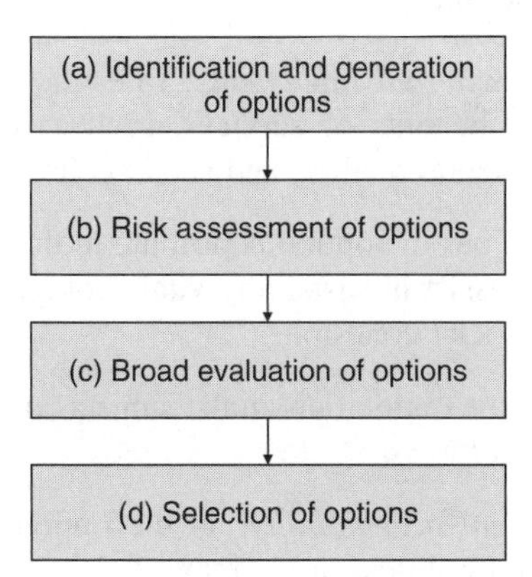

Figure 17.1 Main steps of risk management process (iterations and feedback loops ignored).

(a) **Identification and generation of options**, which reflect, *inter alia*:

- technical standards and limits that prescribe, for example, the permissible threshold of concentrations, emissions, take-up or other measures of exposure;
- performance standards for technological and chemical processes, such as minimum temperatures in waste incinerators;

- technical prescriptions referring to the blockage of exposure (e.g. via protective clothing) or the improvement of resilience (e.g. via immunization or earthquake-tolerant structures);
- governmental economic incentives, including taxation, duties, subsidies and certification schemes;
- third party incentives, that is, private monetary or in-kind incentives;
- compensation schemes (monetary or in kind);
- insurance and liability;
- co-operative and informative options, ranging from voluntary agreements to labelling and education programmes.

All these options can be used individually or in various combinations to meet defined objectives.

(b) **Assessment of options**, including risk assessments and cost–benefit analyses. Each of the options will have desired and unintended consequences. Examples of criteria to be considered are as follows:

- *Effectiveness*: Does the option achieve the desired effect?
- *Efficiency*: Does the option achieve the desired effect with the least resource consumption?
- *Minimization of external side effects*: Does the option infringe on other valuable goods, benefits or services, such as competitiveness, public health, environmental quality, and social cohesion?
- *Sustainability*: Does the option contribute to the overall goal of sustainability? Does it assist in sustaining vital ecological functions, economic prosperity and social cohesion?
- *Fairness*: Does the option burden the subjects of regulation in a fair and equitable manner?
- *Political and legal acceptability*: Is the option compatible with legal requirements and political programmes?
- *Ethical acceptability*: Is the option morally acceptable?
- *Public acceptance*: Will those individuals who are affected by it accept the option? Are there cultural preferences or symbolic connotations that have a strong influence on how the risks are perceived?

Measuring options against these criteria may give rise to conflicting messages and results. Many measures that prove to be effective may turn out to be inefficient or unfair to those who will be burdened. Other measures may be sustainable but not acceptable to the public or important stakeholders.

(c) **A broad evaluation and judgment of options** (also referred to as a management review and judgement, see Chapter 18): This step integrates the evidence as to how the options perform with a value judgement concerning the relative weight which should be assigned to different concerns. Ideally, the evidence should come from experts and the relative weights from decision-makers. In practical risk management, options are evaluated in close co-operation between both experts and decision-makers.

(d) **Selection of options**: Once the different options have been evaluated, a decision has to be made as to which options are selected and which are rejected. In most cases trade-offs have to be made between different concerns.

These steps are in line with the basic model used by decision theory (Morgan, 1990; Hammond *et al.*, 1999). This process model is simple and ignores iterations and feedback loops. For example, selection of options could mean that some options are selected for further assessments.

The role of risk reduction in risk management

The above examples and the general review of risk management clearly show that risk management is not about an isolated risk reduction process but about balancing different concerns, for example costs and improved safety for people and the environment. Risk appetite is required to create value in our society.

Recall the three situations introduced above:

(i) risk linked to the realization of the activity

(ii) risk linked to constraints such as concept selection and technology

(iii) risk linked to 'free variables' within the constraints of (i) and (ii). These free variables are often related to human and organizational factors.

Risk management in case (i) is about balancing different concerns, and risk reduction is normally not the main issue. If it had been, few activities would be realized, as demonstrated by the examples in the introduction to this chapter.

Risk reduction is an objective in case of (ii) and (iii), but there could be several objectives and they could be in conflict. This is illustrated by the adoption of the ALARP principle which states that the risk should be reduced to a level that is as low as reasonably practicable. A risk-reducing measure should be implemented provided it cannot be demonstrated that the costs are grossly disproportionate relative to the gains obtained (Health and Safety Executive, 2001). This principle is a way of implementing the ambition of reducing risk to protect people, the environment and assets. It is acknowledged that risk reduction must be balanced against other concerns, although risk reduction is a goal. The principle makes a stand in favour of safety, as a measure should be implemented unless it cannot be demonstrated that the costs are in gross disproportion to the benefits gained.

To verify ALARP, procedures mainly based on engineering judgements and codes are used, but also traditional cost–benefit analyses and cost-effectiveness analyses (refer to Chapter 18).

One way of assessing 'gross disproportion' is outlined below (Aven and Vinnem, 2007):

- Perform a crude analysis of the benefits and burdens of the various alternatives, addressing attributes related to feasibility, conformance with good practice, economy, strategy considerations, risk, social responsibility, and so on. The analysis would typically be qualitative and its conclusions summarized in a matrix with performance shown by a simple categorization system such as very positive, positive, neutral, negative, very negative. From this crude analysis a decision can be made to eliminate some alternatives and include new ones for further detailing and analysis. Frequently, such crude analyses give the necessary platform for choosing one appropriate alternative. When considering a set of possible risk-reducing measures, a qualitative analysis in many cases provides a sufficient basis for identifying which measures to implement, as these measures are in accordance with good engineering or with good operational practice. Also many measures can quickly be eliminated as the qualitative analysis reveals that the burdens are much more dominant than the benefits.
- From this crude analysis the need for further analyses is determined, to give a better basis for concluding which alternative(s) to choose. This may include various types of risk analyses.
- Other types of analyses may be conducted to assess costs, for example, and indices such as expected cost per expected number of saved lives could be computed to provide information about the effectiveness of a risk-reducing measure or compare various alternatives. The expected net present value may also be computed when found appropriate. Sensitivity analyses should be performed to see the effects of varying key parameters. Often the conclusions are rather straightforward when calculating indices such as the expected cost per expected number of saved lives. If there is no clear conclusion about gross disproportion (the costs are not so large), then these measures and alternatives are clear candidates for implementation. Clearly, if a risk-reducing measure has a positive expected net present value, it should be implemented. Crude calculations of expected net present values, ignoring difficult judgements about valuation of possible loss of lives and damage to the environment, will often be sufficient to conclude whether this criterion could justify the implementation of a measure.
- An assessment of uncertainties in the underlying phenomena and processes is carried out. Which factors can yield surprising outcomes with respect to the calculated probabilities and expected values? Where are the gaps in

knowledge? What critical assumptions have been made? Are there areas where there is substantial disagreement among experts? What are the vulnerabilities of the system?

- An analysis of manageability takes place. To what extent is it possible to control and reduce the uncertainties and thereby arrive at the desired outcome? Some risks are more manageable than others in the sense that there is a greater potential to reduce risk. An alternative can have a relatively large calculated risk under certain conditions, but the manageability could be good and consequently the result could be a far better outcome than expected.
- An analysis of other factors, such as risk perception and reputation, should be carried out whenever relevant, although it may be difficult to describe how these factors would affect the standard indices used in economic and risk analysis to measure performance.
- A total evaluation of the results of the analyses should be performed, to summarize the pros and cons of the various alternatives, where considerations of the constraints and limitations of the analyses are also taken into account.

Note that such assessments are not necessarily limited to the ALARP processes. The above process can also be used in other contexts where decisions are to be made under uncertainty.

ALARP assessments require that appropriate risk-reducing measures are generated. Suggestions for measures always arise in a risk analysis context, but often a systematic approach for the generation of these is lacking. In many cases, the measures also lack ambitions. They bring about only small changes in the risk picture. A possible way to approach this problem is to apply the following principles:

- On the basis of existing solutions (base case), identify measures that can reduce the risk by, for example, 10%, 50% and 90%.
- Specify solutions and measures that can contribute to reaching these levels.

The solutions and measures must then be assessed before making a decision on possible implementation.

A potential strategy for the assessment of a measure, if the analysis based on expected present value or expected cost per expected number of lives saved has not produced any clear recommendation, can be that the measure be implemented if several of the following questions give a yes answer:

- Is there a relatively high personnel risk or environmental risk?
- Are there considerable uncertainties (related to phenomena, consequences, conditions) which the measure will reduce?

- Does the measure significantly increase manageability? High competence among the personnel can give increased assurance that satisfactory outcomes will be reached, for example fewer leakages.
- Does the measure contribute to obtaining a more robust solution?
- Is the measure based on the best available technology?
- Are there unsolved problem areas related to personal safety and/or work environment?
- Are there possible areas where there is conflict between these two aspects?
- Is there a need for strategic considerations?

We refer to Aven and Vinnem (2007). The authorities may also implement other requirements to ensure that people and the environment are protected against hazards and threats. All industries are regulated, and specific requirements may be set to safety barriers, emergency preparedness, and so on. See Chapter 19.

We will discuss the decision process, in particular related to (ii), in more detail in the coming chapter. Concerning (iii), the issue is, to a large extent, about operational measures, for example related to procedures for carrying out the activity, competence and organizational aspects, as well as emergency preparedness measures. Cost may be an issue, but often the decision about implementing the measure is more about believing that the measure will have a positive effect on risk than about costs. If it is strongly believed that a procedure will improve safety, it will be implemented.

Summary

Risk management may be defined as all measures and activities carried out to manage risk. Risk management deals with balancing the conflicts inherent in exploring opportunities on the one hand, and avoiding losses, accidents and disasters on the other. Risk management relates to all activities, conditions and events that can affect the organization, and its ability to reach its goals and vision. It is acknowledged that risk cannot be eliminated but must be managed. To create value, risks must be taken. Risk reduction is consequently not a goal in itself. Given that an activity is to be carried out, however, we will implement risk-reducing measures to ensure desirable outcomes, but there will always be trade-offs between different concerns. A measure may reduce risk but could lead to a less satisfactory solution with respect to other concerns, such as costs. To protect human lives and the environment, the society implements requirements, for example the ALARP principle, ensuring that risk-reducing processes are implemented.

References

AS/NZS (2004) *Risk Management Standard*, AS/NZS 4360: 2004, Jointly published by Standards Australia International, Ltd, Sydney and Standards New Zealand, Wellington.

Aven, T. (2008) *Risk Analysis*, John Wiley & Sons, Ltd, Chichester.

Aven, T. and Renn, O. (2009) The role of quantitative risk assessments for characterizing risk and uncertainty and delineating appropriate risk management options, with special emphasis on terrorism risk. *Risk Analysis*, **29**, 587–600.

Aven, T. and Vinnem, J.E. (2007) *Risk Management*, Springer-Verlag, New York.

CNN (2001) Man on the moon: Kennedy speech ignited the dream. http://archives.cnn.com/2001/TECH/space/05/25/kennedy.moon/ (accessed 25 May).

COSO (2004) Enterprise Risk Management Framework. Committee of Sponsoring Organizations of the Treadway Commission.

Hammond, J., Keeney, R. and Raiffa, H. (1999) *Smart Choices: A Practical Guide to Making Better Decisions*, Harvard Business School Press, Cambridge, MA.

Health and Safety Executive (2001) *Reducing Risk, Protecting People*, HSE Books.

IRGC (2005) Risk Governance: Towards an Integrative Approach, White Paper No. 1, written by Renn O. with an Annex by P. Graham P. (International Risk Governance Council, Geneva).

ISO (2008) *Risk Management – General Guidelines for Principles and Implementation of Risk Management*, Preliminary version.

Jorion, P. (2007) *Value at Risk*, 3rd edn, McGraw-Hill, New York.

Morgan, M.G. (1990) Choosing and managing technology-induced risk, in T.S. Glickman and M. Gough (eds), *Readings in Risk*, Resources for the Future, Washington, DC, pp. 17–28.

Further reading

COSO (2004) Enterprise Risk Management Framework. Committee of Sponsoring Organizations (COSO).

18

Decision-making under uncertainty should be based on science (analysis)

As noted in the previous chapter, risk management and decision-making under uncertainty are often split into four tasks:

1. Identification and generation of options
2. Assessment of these options, using, for example, risk assessments and cost–benefit analyses
3. A broad evaluation and judgement of options (also referred to as a management review and judgement)
4. Decision: selection of options.

These tasks are based on the following pre-studies and definitions.

- Decision situation and the stakeholders (interested parties):
  - What is the decision situation?
  - What are the boundary conditions?
  - Who is affected by the decision?
  - Who will make the decision?
  - What strategies are to be used to reach a decision?
- Goals and preferences:
  - What do the various interested parties want?

Misconceptions of Risk T. Aven

— How to weigh the pros and cons?

— How to express the performance of the various alternatives?

A crude model for decision-making based on the above elements is presented in Figure 18.1.

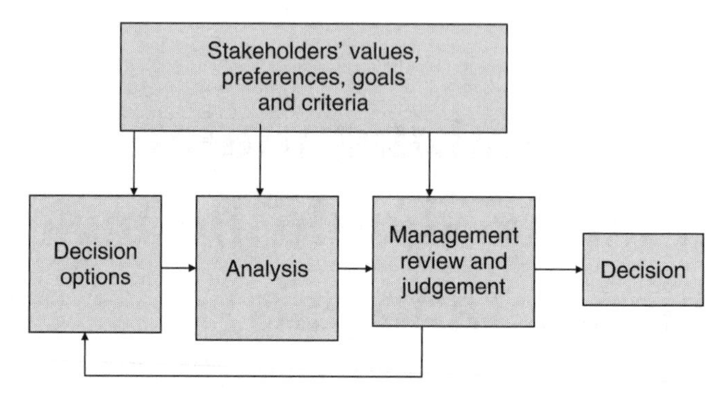

Figure 18.1 A model for decision-making under uncertainty (Aven 2003).

The decision support produced by the analyses must be reviewed by the decision-maker prior to making the decision. In general, there is a leap from the decision-making basis to the decision itself. It is necessary to review the decision-making basis: What is the background information of the analyses? What are the assumptions and suppositions made? The results from the analysis must be assessed in the light of factors such as (Aven, 2008):

- which decision-making options are being analysed?
- which performance measures are being assessed?
- the fact that the analyses represent judgements – to a large extent carried out by experts
- difficulties in determining the advantages and disadvantages of the different alternatives
- the fact that the analysis results are based on models that are simplifications of the real world and real-world phenomena.

The point is that the decision-making basis will seldom be in a format that provides all the answers that are important to the decision-maker. Perhaps the analysis did not take into consideration what the various options/measures mean for the reputation of the enterprise, but this is obviously a condition that is of critical importance for the organization. The review and judgement must also cover this aspect.

The model is based on a clear distinction between analysis and management. Professional analysts provide decision support, not the decision itself. Decision-making in the face of uncertainties and risk is risk-informed, not risk-based (Apostolakis, 2004). As noted in Chapter 6, the need to make a distinction between analysis and values has been discussed thoroughly in the literature, in particular among social scientists. See, for example, Renn (2008), Rosa (1998) and Shrader-Frechette (1991).

However, many experts and managers adopt a perspective for decision-making under uncertainty that is more risk-based than risk-informed. Their goal is to strengthen the analysis (scientific) part and reduce the management (non-scientific) part. Only in this way can rational decisions be made. Their perspective will be presented and discussed in more detail in the following.

A perspective based on science (analysis)

This perspective is based on two main pillars:

1. Risk is an objective property of the activity studied, and the aim of the risk assessment is to estimate this risk such that a rational decision can be made about whether the risk is acceptable or not.
2. If the risk is acceptable, options can be compared and prioritized using cost–benefit analyses where the expected net present value, E[NPV], is calculated.

Consider an investment in a project for the development of a new product. If $E[NPV] > 0$, the project should be realized provided that the risk related to extreme negative outcomes is sufficiently small. The main concern in this example is the possible occurrence of technical problems leading to large development costs. A requirement like this may be formulated: the probability that such technical problems occur should not exceed 5%.

A risk analysis is then conducted and the probability of the event of interest is estimated to be 2%. The conclusion is that the risk is considered acceptable. Provided that the calculations of the E[NPV] are positive, the project should be realized. Figures 18.2 and 18.3 illustrate these procedures.

Within investment analysis and risk analysis in finance, risk indices such as the value-at-risk (VaR) and the tail-value-at-risk (TVaR) are commonly used; see Chapter 4. Instead of $P < p_0$, we could have formulated requirements based on, for example VaR. The following discussion still applies.

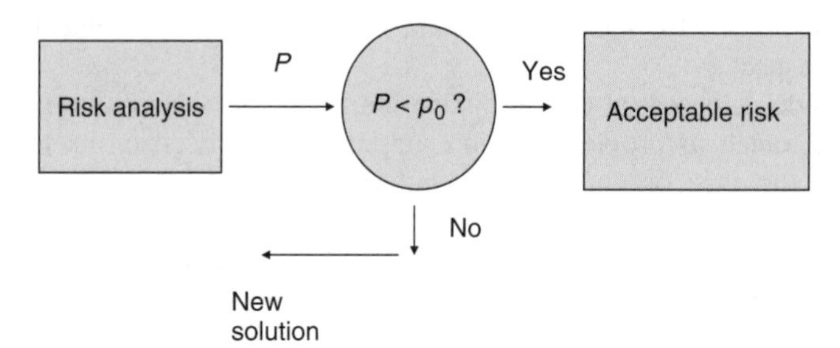

Figure 18.2 Illustration of how the risk analysis is used to determine whether risk is acceptable or not. The criterion is that the probability P for a specific event is less than a specified p_0.

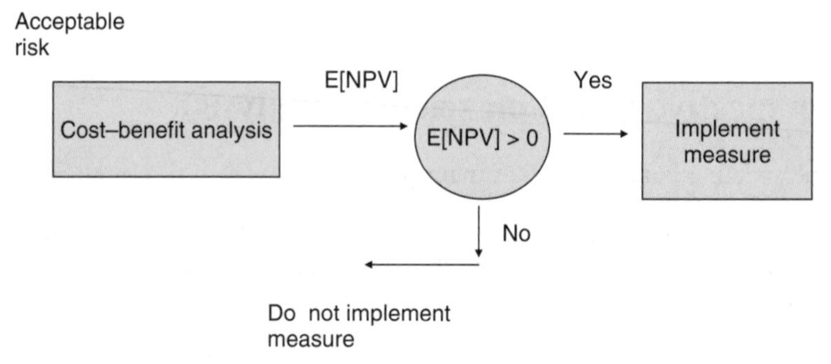

Figure 18.3 Illustration of how the cost–benefit analysis is used to decide whether a measure or project should be realized or not. The criterion is that the expected net present value, E[NPV], is positive.

Critique of this approach

The approach described by Figures 18.2 and 18.3 seems appealing. However, a closer look reveals several problems. The numbers have been given an authority which cannot in general be justified: the estimates are subject to large uncertainties. Under different assumptions and/or the use of other analysts, the results could have been different. And we may also question the principles and ideas used. A solution may be rejected if the probability estimate is too high, but this solution may be extremely attractive when looking at other attributes. It may be worth taking the risks. This is the key issue. It may not be wise to define absolute constraints, here $P < p_0$, before the benefits and costs of all attributes have been assessed. The problem is the same as the one you face when looking for a wife or a new car. If, early in the process, you disregard all cars exceeding $20 000 or determine that your future wife should not be blond, your option space is strongly reduced and very good candidates from an overall perspective may be overlooked. For an organization, an overall best solution is what we should look for.

In addition, one may challenge the use of the E[NPV] criterion. The risk criterion may be satisfied, but stronger weight on the risk may be desirable than that which is obtained by using the E[NPV]. An expected value gives, as we have discussed earlier (see Chapter 1), little weight to the extreme events – one multiplies small probabilities and large losses. The need to emphasize the risk and uncertainty dimension beyond the expected values is particularly relevant when considering accidents and threats affecting human lives and the environment. The authorities may, for example introduce the ALARP principle (refer to Chapter 17), as is common in many countries and industries. The ALARP principle takes a stand in favour of safety, in the sense that a measure should be implemented unless it cannot be documented that the costs are in gross disproportion to the benefits gained. The burden of proof is reversed. The E[NPV] is not suitable to reflect such a criterion (Aven and Vinnem, 2007).

In E[NPV] calculations, the discount rate is risk-adjusted, but only systematic risk (uncertainty) is taken into account, that is, risk that affects the whole market and cannot be diversified. See Chapter 4 and, for example, Damodaran (2002). Hence, the use of E[NPV] means that extreme events are not given more weight than the product of probability and consequence.

The above approach, which is based on risk assessments and cost–benefit analysis, has been strongly criticized by many analysts and researchers. O'Brien (2000) is a good example. Her focus is the use of risk assessment to defend activities that could harm the environment and human health, in other words to defend judgements about acceptable risk. O'Brien (2000) gives a number of examples, most related to toxic chemicals. Her point is clear. Risk assessments generally serve the interests of business (i), as well as government agencies (ii) and many analysts (iii).

(i) Through risk assessments, an industry gets significant legal protection for activities that result in contamination of communities, workers, wildlife and the environment with toxic chemicals. Through risk assessment, industry gets protection for filling streams with sediments, thinning the ozone layer, causing high cancer rates, avoiding cleaning up its own mess, and earth-damaging activities (O'Brien, 2000, p. 102). The risk assessment gives the industry the aura of being scientific. The risk assessments show that the activities are safe, and most of us would agree that it is rational to base our decision-making on science. The complexity of a risk assessment makes it difficult to understand its premises and assumptions if you are not an expert in the field. In a risk assessment there is plenty of room for adjustment of the assumptions and methods to meet the risk acceptance criteria. In the case of large uncertainties in the phenomena and processes studied, the industry takes advantage of the fact that in our society safety and environment-affecting activities and substances are considered innocent until 'proven guilty'. It takes several years to test, for example, whether a certain chemical causes cancer, and the uncertainties and choice of appropriate risk-assessment premises and assumptions allow interminable haggling.

(ii) Risk-assessment processes allow government to hide behind 'rationality' and 'objectivity' as they permit hazardous activities that may harm people and the environment (O'Brien, 2000, p. 106). The focus of the agencies is then more on whether a risk assessment has been carried out according to the rules, than on whether it provides meaningful decision support.

(iii) Risk analysts know that the assessments are often based on selective information, arbitrary assumptions and enormous uncertainties. Nonetheless, they accept that the assessments are used to conclude on risk acceptability.

This critique of risk assessment is strong, and it is supported by many other researchers, see, for example Reid (1992), Stirling (1998, 2007), Renn (1998), Tickner and Kriebel (2006) and Michaels (2008). Reid (1992) argues that the claims of objectivity in risk assessments are simplistic and unrealistic. Risk estimates are subjective, and there is a common tendency to underestimate the uncertainties. The disguised subjectivity of risk assessments is potentially dangerous and open to abuse if it not recognized. According to Stirling (2007), using risk assessment when strong knowledge about the probabilities and outcomes does not exist is irrational, unscientific and potentially misleading. Renn (1998) summarizes critiques drawn from the social sciences over many years and concludes that technical risk analyses represent a narrow framework that should not be the single criterion for risk identification, evaluation and management. Tickner and Kriebel (2006) pp. 53–55 and Michaels (2008) argue along the same lines as O'Brien (2000). Tickner and Kriebel (2006) particularly stress the tendency of decision-makers and agencies not to talk about uncertainties underlying the risk numbers. Acknowledging uncertainty can weaken the authority of the decision-maker and agency, by creating an impression of being unknowledgeable. Precise numbers are used as a façade to cover up what are often political decisions. Michaels (2008) argues that mercenary scientists, including risk analysts, have increasingly shaped and skewed the technical literature, manufactured and magnified scientific uncertainty, and influenced government policy to the advantage of polluters and the manufacturers of dangerous products.

The answer to this critique, according to O'Brien (2000), is to look for an alternative to risk assessments. The critique of risk assessment is to a large extent justified, but to support the decision-making we need to assess risk. The right move forward is not to reject risk assessment, but to improve the tool and its use (Aven, 2009). This seems also to be the conclusion of most researchers in the field. The challenge is how decision-making on risk can be informed by the best available technical and scientific knowledge (see Stirling, 1998, p. 100;

Apostolakis, 2004). Different ways of strengthening the quality of the risk assessments and the associated risk assessment process to meet the above critique have been suggested. Three main success factors are identified and discussed in Aven (2009):

1. The scientific basis of the risk assessments needs to be strengthened. Stakeholders and decision-makers need to improve their understanding of the fundamentals of risk assessment.

2. Risk assessments need to provide a much broader risk picture than is typically the case today. Separate uncertainty analyses should be carried out, extending the traditional probability-based analyses.

3. The cautionary and precautionary principles need to be seen as rational risk management approaches, and their application should, to a large extent, be based on the risk and uncertainty assessments.

Other researchers have presented similar recommendations; see, for example Stirling (1998), 2007) and Renn (1998). In the social scientific risk research community there has been a strong recognition of the need to move away from a search for an 'analytical fix' and towards addressing the social and institutional aspects of the problem. IRGC (2005) and UK Cabinet Office (2002) are examples of risk frameworks based on this recognition. Fischhoff (1995) has summarized seven stages in this retreat away from what might been seen as 'naïve positivism' in the risk debate (Stirling, 1998):

1. All we have to do is get the numbers right.

2. All we have to do is tell them the numbers.

3. All we have to do is explain what we mean by the numbers.

4. All we have to do is show them that they've accepted similar risks in the past.

5. All we have to do is show them that it's a good deal for them.

6. All we have to do is treat them nice.

7. All we have to do is make them partners.

Nonetheless, the risk assessment consultants and the formal decision-making on the regulation of risk remain relatively unaffected by this recognition (see Stirling, 1998, p. 100). There are many reasons for this, but a major factor is certainly the risk assessment tool in itself. The assessments are, in general, lacking a proper foundation, the perspective on uncertainties is too narrow and there is a common misconception that risk tolerability and risk acceptance can be defined through predefined limits and criteria. Improvements on these issues are required (Aven, 2009).

There is a huge body of scientific literature that aims to strengthen the quality of risk assessments. However, most of this research is focused on the development of new and improved risk assessment approaches and methods. Relatively little attention has been devoted to foundational issues. Uncertainty is a frequently occurring topic in risk assessment research but its scope is often very technical. The literature is characterized by some prevailing ways of thinking on how to deal with uncertainties in risk assessments; see, for example Paté-Cornell (1996). We seldom see fundamental discussion about uncertainty in a risk assessment context that extends beyond this thinking. It is acknowledged that risk assessments provide decision support – the decisions are risk-informed and not risk-based (Apostolakis, 2004) – but what are the key aspects to consider outside the realm of the traditional risk assessment results?

Aven (2008, 2009) argues for the need to assess uncertainty factors which are 'hidden' in the background knowledge of the assigned probabilities and expected values. The results of the risk assessments must be extended beyond the probabilistic world. In line with this thinking we must also redefine the concept of risk: uncertainty should replace probability in the definition of risk. The uncertainty factors addressed are then linked to the decision-making process and the use of the cautionary and precautionary principles. The uncertainties in most cases lead to the use of the cautionary principle, not the precautionary principle as discussed in the coming chapter.

Conditions for obtaining improved risk assessments and risk assessment processes

A risk assessment must have a scientific basis, which clarifies the axioms, interpretations and measurement procedures of the representations of uncertainties, that is, the probabilities and risk indices (Bedford and Cooke, 2001). The present situation is, however, not acceptable, as illustrated by the following two examples:

(i) The risk assessment typically has a focus on the estimation of some probabilistic parameters, for example the probability p of an attack. A meaningful interpretation of these parameters is, however, often lacking; see Chapter 14. It should be a basic requirement of the assessment that the quantities introduced are properly defined and are meaningful representations of the activity or system being studied. If we are to assess uncertainties of average performance of quantities of such populations, it is essential that we understand what they mean.

(ii) If we challenge a number of risk professionals on what they mean by the term 'model uncertainty' we are likely to get as many answers as experts. Models play an important role in risk assessments, but there is a lack of understanding of this role. If model uncertainty is a relevant term, we need to address the uncertainties.

These two examples show that the scientific basis of risk assessment can be questioned. The literature includes many contributions that try to clarify key concepts and obtain order and structure (Bedford and Cooke, 2001; Haimes, 2004; Aven, 2003; Vose, 2008). However, there are many open questions related to the fundamental issues in risk assessment and there are different views in the scientific risk community: some researchers and analysts would introduce frequentist probabilities of an attack, others reject their existence. Some acknowledge that model uncertainty could be a significant problem in risk analysis, but find it hard to prescribe a method for dealing with it in practice. And so on. We could make a long list of challenges that the risk community faces. It is unrealistic to expect broad consensus on how to deal with all these issues, but we should sharpen the focus on the foundational issues and the scientific platform of the assessments. Aven (2009) highlights several conditions which are believed to be critical for obtaining the desired strengthening of the scientific basis of the assessments. Some of these conditions are discussed below.

It is not possible to obtain a proper use of risk assessments in a decision-making context unless the key stakeholders have a solid understanding of what a risk assessment is. The risk analysts' competence is essential in this regard. They need to establish a proper scientific basis for their assessment, as stressed above. This requires that the analysts are professional in the field of risk and uncertainty. Now we often see risk analysts with a poor background in the fundamentals of risk assessment. It is not sufficient to be a statistician or an engineer to act as a professional risk analyst. Risk assessment is a discipline in its own right and requires education and training in topics like risk and probability concepts, risk analysis methods, uncertainty analysis, risk characterizations and risk communication. The extent to which stakeholders – including decision-makers and third parties – understand risk assessments and their results depends very much on the professional risk analysts' ability to communicate the risk picture. Risk professionals need to be extremely clear as to the scope, as well as the boundaries and limitations, of the assessments.

A risk assessment produces risk numbers in the form of probabilities and expected values. But this risk picture is, in general, too narrow: uncertainties are not properly reflected. Uncertainty assessments extending beyond the probabilities and expected values should be included. The uncertainty assessments should not be restricted to standard probabilistic analysis, as this analysis could hide important uncertainty factors. For example, you may assign a probability of fatalities occurring on an offshore installation based on the assumption that the installation structure will withstand a certain accidental load. In real life, however, the structure could fail at a lower load level. The probability did not reflect this uncertainty. Or you may assign a low probability of health problems occurring as a result of some new chemicals, but these probabilities could produce poor predictions of the actual number of people that experience such problems. Refer to Chapters 2 and 12.

The broad risk description should cover risk numbers and sensitivities as well as uncertainty factors. Business may not benefit from such a risk description, but risk analysts and the risk analysis community should be clear on the need to describe risk in this way. A narrow perspective on risk cannot be justified.

The need for additional uncertainty assessments goes beyond confidence and credibility intervals as in traditional risk assessments. A confidence interval just produces a measure of the variation in the data. The relevance of the data is not taken into account. Hence, for most risk assessments, the uncertainty is not adequately reported if confidence intervals are used.

A more extensive uncertainty assessment is carried out when adopting the probability of frequency approach which was introduced in Chapter 2 (Kaplan and Garrick, 1981; Kaplan, 1991; Apostolakis, 1990; Paté-Cornell, 1996). Here second-order probabilities (knowledge-based or subjective probabilities) are used to describe the assessors' uncertainty about the frequentist probabilities (or the chances in a Bayesian setting). The ambition of the probability of frequency approach is to express the epistemic uncertainties of the probability p of an attack (say) and take into account all relevant factors causing uncertainties. The analysis may produce a 90% credibility interval for p, according to which the analyst is 90% confident that p lies in the interval $[a, b]$.

In practice, however, it is difficult to perform a complete uncertainty analysis following this approach. In theory, an uncertainty distribution on the total model and parameter space should be established, which is impossible to do. So in applications only a few marginal distributions on some selected parameters are normally specified and, therefore, the uncertainty distributions on the output probabilities merely reflect some aspects of the uncertainty. This makes it difficult to interpret the uncertainties produced.

Hence, if the assessment produces a 90% credibility interval [0.001, 0.01] for the parameter, it may not be clear how this interval should be interpreted as it does not reflect the uncertainties of all input parameters. In more general terms, we face the problem that this interval represents an uncertainty assessment based on a background knowledge (including assumptions, suppositions, models) and this background knowledge could conceal or camouflage uncertainties. The point we are making is that the probabilistic numbers represent someone's assessments – based on certain background knowledge. This background knowledge could be poor, and this aspect of uncertainty is often not properly reflected by the assessments. But it certainly has to be taken into account in the risk management. This is the motivation for addressing the uncertainty factors which extend beyond the probabilistic world. We refer to Chapters 2 and 12 for further discussion of these issues.

Example. Cash depot case

A company, NOKAS, performs both Norwegian central bank tasks and other cash processing services for private banks. In April 2004 NOKAS was just about to move into new premises in the city of Stavanger when a robbery at one of the existing facilities took place. Approximately $8 million were taken, and one policeman was killed during the attack. This brutal robbery gave rise to huge concern among the neighbours of the new NOKAS facility which is situated in a residential area, and located only 8 m from a kindergarten (Vatn, 2007; Aven, 2008). The risk decision problem can be formulated as follows: is the risk for the neighbours acceptable? Should the NOKAS facility be relocated?

A risk assessment was performed and, based on this assessment, it was concluded by NOKAS and the bureaucrats of the municipality to conclude that the risk is acceptable, provided that some risk-reducing measures were implemented. Thus, relocation of the NOKAS facility was not required.

At an early stage of the assessment process, NOKAS hired a consultancy company and together they established some acceptance limits for third parties' risks (including risk for the children and the neighbours). Risk assessments were then carried out and as the risk numbers produced were below the criteria specified, it was concluded that the risk was acceptable.

The assessment was based on a traditional probability analysis. The huge uncertainties were not communicated. The probability analysis in this case had to be based on strong assumptions, for example concerning the number of attacks and the manner in which an attack might take place. But these quantities are subject to large uncertainties. A possible scenario is that an attack will occur in an innovative, surprising and brutal way. The analysts may assess the likelihood related to these factors, but as the uncertainties are so large they constitute a part of the risk picture. Using the predefined probabilistic acceptance limits, the issue of risk acceptance has been reduced to one in the probabilistic world of the consultant. The underlying uncertainties have been camouflaged for the other stakeholders and the decision-makers, as the results, to a large extent, are presented as probabilities. Risk is reduced to the probabilities and the narrow number-crunching exercise shows that the risk tolerability limits are met. Yes, the risk assessments serve the interests of the business.

The uncertainties were not given sufficient weight to justify the removal of the NOKAS facility, and the decision was made not to remove the facility. We may just speculate whether a broader risk characterization would have changed the decision. It could have, as the uncertainties related to future attacks and the way a possible attack will take place would have been given more weight in the risk picture presented to the politicians. The administration and the bureaucracy could not have provided as clear recommendations as they did. The politicians would, thus, have seen a report and a message with no clear recommendations on what to do. The message from the analysts would have been that the task of balancing different concerns and giving weight to the uncertainties is a management (here political) responsibility.

One may also question the extent to which the politicians would have appreciated a more informative and reflective perspective. There will clearly be different opinions on such a perspective among the politicians. Some will be sceptical because they must do more independent evaluations and thus accept greater responsibility. Others will, hopefully, say that this is the correct path. □

An important point made by O'Brien (2000) is that risk assessments give the industry the aura of being scientific, and they can refer to risk being acceptable according to the results of the assessments. The above conditions for improving the quality of the risk assessment would give a more nuanced perspective on what 'scientific' means and make it impossible to use the assessment to conclude on risk acceptability on the basis of the risk assessment alone. The subjectivity of the risk assessment results is acknowledged. There are uncertainties present and the weight given to them is not a scientific issue, but a management (political) issue. In the NOKAS case this would have resulted in a different process as the consultant could not have concluded on risk acceptability and more responsibility would have been placed on the politicians, as discussed above.

Another issue raised by O'Brien (2000) is the industry's possibility of being considered innocent until 'proven guilty'. Improving the quality of the risk assessments and the risk assessment processes will not solve this problem. This issue is a management (political) issue: how to balance different concerns and, in particular, how to weight the cautionary and precautionary principles. It extends further than the risk assessments. However, the above conditions for improving the quality of the risk assessments could lead governments and government agencies to rely less on the risk assessments being able to come up with clear answers. In case of considerable uncertainties, the traditional risk assessments do not offer a rigorous and robust basis for decision-making (see Stirling, 2007). The importance of the cautionary and precautionary principles would be increased, as uncertainties beyond the calculated probabilities need to be taken into account. This certainly creates a regulatory challenge for the government agencies – it requires more agency involvement. The agencies cannot hide behind 'rationality' and 'objectivity', checking that a risk assessment has been carried out according to the rules, and accepting that the risk assessment is used to prove that the activity is 'safe'. The weight given to the uncertainties is not for the business alone to determine. It would not be possible to refer to the assessment being conducted without addressing the uncertainties, as well as its limitations and constraints.

Cost–benefit analysis based on expected net present value and other types of criteria

The above discussion has focused on risk assessments. However, it is also, to a large extent, applicable to the use of expected net present calculations and the expected utility: there is a need to look beyond the probabilities and expected values; see Chapters 2 and 6.

A number of decision analysis methods are available to management faced with the evaluation of decision alternatives under uncertainty. The methods include expected utility analysis, cost–benefit analysis, cost-effectiveness analysis and multi-attribute analysis. According to expected utility theory, rational behaviour for a single decision-maker is equivalent to maximizing expected utility, that is, to choosing the decision alternative with the highest expected utility;

see Chapter 6. Its logical basis gives the expected utility theory a strong position as a normative theory, but the utility concept is difficult to implement in practical decision-making. The practical solution to this problem is often to use cost–benefit analyses based on E[NPV] calculations. To calculate the NPV, the relevant cash flows (the movement of money into and out of the business) are specified and the time value of money is taken into account by discounting future cash flows. The formula used to calculate the NPV is

$$\mathrm{NPV} = \sum_{t=0}^{T} \frac{Y_t}{(1 + r_t)^t}, \tag{18.1}$$

where Y_t is equal to the cash flow at year t, T is the time period considered (in years) and r_t is the required rate of return, or the discount rate, at year t. The terms 'cost of capital' and 'opportunity cost' are also used for r_t. As these terms indicate, r_t represents the investor's cost of not employing the capital in alternative investments. When considering projects where the cash flows are known in advance, the rate of return associated with other risk-free investments, such as bank deposits, forms the basis for the discount rate to be used in the NPV calculations. When the cash flows are uncertain, which is usually the case, they are represented by their expected values, and the E[NPV] is obtained. The discount rate used to calculate the E[NPV] is adjusted on the basis of the capital asset pricing model (CAPM) to compensate for uncertainties (risk). See, for example Levy and Sarnat (1994) and Varian (1999); refer also to Chapter 4. However, not all types of uncertainties are considered relevant when determining the magnitude of the risk-adjusted discount rate, as shown by the portfolio theory; the portfolio theory justifies the ignorance of unsystematic risk and states that only systematic risk associated with a project is relevant when taking a portfolio perspective. Systematic risk relates to general market movements, for example the risk associated with political events, whereas unsystematic risk relates to project-specific uncertainties, for example the risk related to accidents.

The method requires transformation of goods into monetary values, for example the value of a 'statistical life'. What is the maximum amount the decision-maker is willing to pay for reducing the expected number of fatalities by 1? Typical numbers for the value of statistical life used in cost–benefit analysis are \$1–10 million. The Ministry of Finance in Norway has arrived at a value of approximately \$3 million. For official cost–benefit analyses, the Ministry of Finance recommends use of a value of this order of magnitude.

The cost–benefit analysis was originally developed for the evaluation of public policy issues, and based on judgements as to how much society was willing to pay to obtain a benefit. Our focus, however, is on the use of cost–benefit analysis as a decision-maker tool and reflecting its willingness to pay.

In practice, cost-effectiveness indices such as the expected cost per expected number of saved lives (often referred to as the implied cost of averting a statistical fatality, ICAF), are often used instead of full cost–benefit analyses. If a measure costs \$2 million and the risk analysis shows that the measure will reduce the expected number of fatalities by 0.1, then the ICAF is equal to $\$2/0.1 = \$20$ million. By comparing this number with reference values, we can assess the effectiveness of the measure.

Example. Cash depot case continued

A number of risk-reducing measures were suggested, including:

- relocation of the NOKAS facility
- relocation of the kindergarten
- erection of a wall between the NOKAS facility and the kindergarten
- covering the NOKAS facility with panels
- review of the emergency preparedness procedures for the kindergarten.

We will take a closer look at one of these measures below:

Relocation of the NOKAS facility

The measure that obviously has the greatest risk-reducing effect for third parties is to relocate the NOKAS facility. If the facility is relocated to an area that is zoned exclusively for commercial activity, using an appropriate location plan, there will be few persons (third parties) exposed compared to the threat scenarios identified in the risk analysis. The analysis group's assessment is that the risk then would be practically eliminated. The cost of relocating the NOKAS facility is calculated to be \$10 million. For a period of 30 years, the expected number of saved lives will be 0.03 (calculated on the basis of 0.001 per year) which means that the cost per expected number of lives saved will be $\$10/0.03 = \$330$ million. This is a very large amount, and normally one would conclude that the cost is disproportionate compared to the benefits gained.

The argument here, however, is a traditional cost–benefit analysis and, as discussed above, this approach is not very well suited for expressing the benefits of the safety measure. The risk reduction is greater than the changes in the computed expected value. How much is it worth to remove this risk? The question is about assigning weights to various concerns, including risk. It is clear that significant emphasis must be given to the uncertainties (and the cautionary principle, see Chapter 19) if one is to be able to justify this measure. □

These tools, the E[NPV] and the ICAF, have a strong position in industry. They are frequently used to support decision-making in safety and security contexts as well as production assurance (HSE, 2001; ISO, 2008; Bedford and Cooke, 2001). Theoretically there is also strong support for the use of expected value-based decision criteria. The main justification is the law of large numbers, according to which the average of a number of random quantities can be accurately approximated by the expected value when the number of quantities is high. The portfolio theory plays a similar role in economic theory – it justifies the use of expected values to support decision-making when considering a large number of projects (and ignoring the systematic risk).

However, risk and uncertainty are not adequately taken into account in these tools, and the literature presents a number of modifications. The aim is to better reflect risk aversion: we (i.e. the decision-makers) dislike negative consequences so much that these are given more weight than is justified by reference to the expected value (Levy and Sarnat, 1994). It is acknowledged that we need to look beyond the computed expected values. However, there are lots of ways of extending the traditional approach based on E[NPV] and the ICAF. How should we determine what is the correct or best modification? There needs to be a rationale supporting the approach.

But such a rationale is difficult to find. The extended approaches have a strong element of arbitrariness in the way they are defined, so care has to be shown when using these approaches.

Let us take a closer look at the methods suggested to adjust the E[NPV] and ICAF approaches for risks and uncertainties (EAI, 2006; Hull, 1980; Jonkman *et al.*, 2003; Walls, 2004):

(a) risk-adjusted discount rate

(b) certainty equivalents

(c) downward revision of benefits, upward revision of costs

(d) safety margin

(e) negative safety margin

(f) cut-off periods

(g) distribution of the NPV

(h) risk aversion-adjusted ICAF

(i) integrated risk index.

In the following a short summary of the meaning of these methods is given, based on Aven and Flage (2009).

(a) **Risk-adjusted discount rate.** The E[NPV] is based on a risk-adjusted discount rate, but the adjustment relates to systematic risk only. To take into account the unsystematic risk, we may adjust the rate further upwards. This approach means that less weight is given to benefits and costs in the distant future, the argument being that these are more difficult to predict than cash flows in the near future.

(b) **Certainty equivalents.** In this approach the expected cash flows, $E[Y_t]$, are replaced by their certainty equivalents, that is, the constant value that makes the decision-maker indifferent between the fixed value and the unknown/uncertain cash flow. The certainty equivalent is smaller than the expected cash flow if the decision-maker is risk-averse. The difference between the certainty equivalent and the expected net benefit is often called the risk premium and is the amount the decision-maker would be willing to pay to avoid having to carry the risk and uncertainty of the project (in this case for year t).

Determining the certainty equivalents is difficult in practice, and simplifications are introduced. One approach is based on the introduction of an exponential utility function, which means assuming a constant risk aversion, that is, that the risk premium will not change as a function of the expected net benefit. In this case the certainty equivalent at time t, CE_t, is found by using the formula

$$\mathrm{CE}_t = \mathrm{E}[Y_t] - \frac{\mathrm{Var}[Y_t]}{2R}, \tag{18.2}$$

where R is a non-negative quantity reflecting the risk tolerance of the decision-maker. A low risk tolerance means that CE_t is much smaller than $\mathrm{E}[Y_t]$.

(c) **Downward revision of benefits, upward revision of costs.** The use of certainty equivalents is difficult to carry out in practice and it is sometimes suggested that it be replaced by some ad hoc procedure, reducing the uncertain benefits by, for example, 10% or, alternatively, increasing the uncertain costs by, for example, 10%. Hence, if the expected benefit is 10, the value 9 is used in the overall calculations of the decision criterion. In case of large uncertainties, the percentage should be further decreased (increased).

(d) **Safety margins.** Instead of requiring that the E[NPV] should be positive, we may require that it should be larger than y, where y is a specified number greater than zero. We may interpret y as the desired 'safety margin' for acceptance of the project. The idea is that a larger E[NPV] is required due to the risks and uncertainties involved. Hence, stronger evidence is required to justify the project.

(e) **Negative safety margins.** HSE (2001) sets the ICAF equal to £1 million (hence, the value of a statistical life is set to £1 million). However, for the offshore industry an ICAF of £6 million is considered to be the minimum level, that is, a proportion factor of 6 (HSE, 2009). This value is used in an ALARP context, and defines what is judged as 'grossly disproportionate'. According to the ALARP principle, a risk-reducing measure should be implemented provided it cannot be demonstrated that the costs are grossly disproportionate to the possible benefits gained. Use of the proportion factor 6 is said to account for the potential for multiple fatalities as well as uncertainties. Hence, the base case is that a risk-reducing measure should be implemented, and strong evidence (costs) is required to justify non-implementation.

(f) **Cut-off periods.** Many projects involve large negative net benefits in the early stages and positive net benefits in the later stages. A simple approach requiring strong evidence to justify the project is then to cut off the period of positive net benefit flows after, for example, three or four years. The method implies that net benefits beyond the cut-off period are considered to have a value equal to zero. The approach is conservative in the sense that possible benefits in the distant future are not taken into account.

(g) **Distribution of the NPV.** To see the effects of risks and uncertainties, a natural approach is to compute the probability distribution of the NPV. Then quantiles of the distribution can be presented, for example the probability that the NPV is greater than any number z. The analysis is often carried out using Monte Carlo simulation, assuming independent cash flows. To take dependence into account, the simultaneous distribution of all cash flows is required, which in practice leads to the use of the normal distribution, where it is sufficient to specify the means, variances and correlation coefficients. Instead of determining the full distribution, it is common to focus on the variance and standard deviation of the NPV. Using formula (18.1), it is straightforward to establish a formula for the variance of the NPV based on the variance of the cash flows and the correlation coefficients for pairwise cash flows.

(h) **Risk aversion-adjusted ICAF.** The ICAF expresses the expected cost per expected number of saved lives. To reflect risk aversion, a number of indices have been suggested for adjusting the contribution of the number of fatalities. See Jonkman *et al.* (2003), p. 7. One example of such an index is (Vrijling *et al.* 1995)
$\mathrm{E}[N] + k\sigma(N)$, where N is the number of fatalities, $\sigma(N)$ is the standard deviation of N and k is a risk aversion factor.

(i) **Integrated risk index.** The result of an activity is the occurrence of one of the scenarios $i = 1, 2, \ldots, n$, where scenario i has probability P_i and consequence C_i. Hence, the expected value equals

$$\sum_{i=1}^{n} P_i C_i.$$

Bohnenblust (1998) suggests the use of the more general formula

$$\sum P_i C_i \varphi(C_i)\omega_i,$$

where $\varphi(C_i)$ is a risk-aversion function depending on the consequences C_i, and ω_i is the willingness to pay for measures to prevent scenario i.

We see that if $\varphi = 1$ and $C_i = 1$, this formula reduces to the expected value.

Discussion

All these methods seek to improve the basic E[NPV] and ICAF approaches by taking into account risks and uncertainties. However, they can all easily be criticized. Although arguments are provided to support these methods, their rationale can be questioned. There is a significant element of arbitrariness associated with the methods.

The approach where the discount rate is adjusted (a) seems plausible as the systematic risk is incorporated in the NPV calculations. But how large should the adjustment be? There is a rationale for the systematic risk adjustment – the CAPM model – but such a rationale does not exist for unsystematic risk. It will, in fact, be impossible to find such a rationale as the calculations are based on expected cash flows, which to a large extent ignore the unsystematic uncertainties. The method is clearly not appropriate if the main risks and uncertainties stem from the early periods. This could be the case if, for example, the costs of a project are highly uncertain, while the future benefits are less uncertain. The method is conceptually unappealing as it blends accounting for the true time preference and for risk aversion (EAI, 2006; Treasury Board of Canada, 1998; Chapter 8).

A traditional NPV analysis is not suitable for reflecting the operating flexibility associated with a project, such as the possibility (option) to make or revise decisions in the future (e.g. defer, expand or abandon the project). The same holds for the strategic (option) value of the interdependence between the project and future or follow-up investments. However, methods exist where this type of management flexibility is explicitly taken into account in the analysis, leading to a modified NPV (Trigeorgis and Mason, 2001). This modified NPV reflects unsystematic risk to a larger extent than traditional NPV analyses, but the unsystematic risk is still only reflected through the expected value calculations.

The use of certainty equivalents (b) means an expected utility approach for the cash flows seen in isolation. To simplify, assume that the NPV relates to two years only, and the cash flows are Y_0 and Y_1. In this approach, the uncertain cash flows are replaced by their certainty equivalents, CE_0 and CE_1 respectively, which means that the uncertain cash flow Y_i is seen as equivalent to having the money CE_i with certainty, $i = 0, 1$. The specification of certainty equivalents is not straightforward, as already mentioned. However, the important point here is not this specification problem, but the fact that this procedure does not necessarily reflect the decision-maker's preferences. If we ignore the discounting for a moment, the utility function of the cash flows Y_0 and Y_1 is not, in general, given by the sum of the individual utility functions. By introducing certainty equivalents on a yearly basis, we take uncertainties into account, but the way we do it has not been justified (Aven and Abrahamsen, 2007).

This critique also applies to formula (18.2), and in addition we have to justify an exponential utility function.

We understand the motivation for method (c), namely to simplify the certainty equivalent approach. However, it is not possible to justify a fixed percentage reduction (increase) of the benefits (costs).

Introducing a safety margin (d) is a way of taking into account risk and uncertainties, the consequences being that fewer projects are accepted than when using the standard approach. If the projects are risk-reduction projects, the safety margin approach makes no sense. For such projects the negative safety margin method (e), for example based on the ALARP principle, is more appropriate. A measure should be implemented unless it can be demonstrated that the costs are in gross disproportion to the benefits gained. Hence, a measure could be justified even in the case of high ICAF values. However, for both the safety margin approach and the negative safety margin approach, risk and uncertainty are restricted to changes in expected values, which is obviously far too narrow a perspective on risk and uncertainty. For example, if new technology is introduced, the expected value may not be significantly disturbed, but the risks and uncertainties increase considerably.

The cut-off method (f) can be viewed as a way of performing sensitivity analysis. What are the effects on the performance indices of ignoring the benefits from a specified year?

The approach where the probability distribution of the NPV is expressed (g), to a large extent, describes the risks and uncertainties. However, the full distribution is difficult to establish as it requires very detailed input information concerning the distribution of the cash flows. Restricting attention to the variance and the standard deviation significantly reduces the analysis, but it is still difficult to establish all the necessary input data.

Many attempts have been made to adjust the contribution from the number of fatalities in order to reflect risk aversion, but it is difficult to argue in favour of one adjustment compared to another. The arbitrariness is large. Both methods (h) and (i) provide adjusted indices reflecting risk aversion, but they lack a rationale.

In view of the above review and discussion, we conclude that all these methods/approaches fail to meet the requirement of incorporating risk and uncertainties. The main points can be summarized as follows:

(i) There is a lack of rationale for all of the adjusted approaches, except the distribution of the NPV (approach (g)). The arbitrariness is large.

(ii) Methods (d) and (e) are based on adjustments of the expected values, which do not adequately reflect risk and uncertainties.

(iii) Risk and uncertainties are represented by probabilities, but probabilities are not perfect tools for expressing risks and uncertainties.

Item (iii) is an important issue. It has been thoroughly discussed throughout this book.

Summary and final remarks

Risk assessments can provide useful decision support, but the assessments are often misused: their results are reported as showing the truth about the risk and being sufficient for drawing conclusions on risk acceptability. Such a practice cannot be justified. The risk assessments and risk assessment process need to be improved. In this chapter we have highlighted conditions that are considered important for obtaining more informative risk assessments and better use of the risk assessments. The main condition is a broad recognition among all stakeholders that risk is more than computed probabilities and expected values, and that the numbers alone cannot and should not be used for mechanical decisions on risk acceptability. Uncertainty is the main component of risk, not probability. The implications would be a more humble attitude to knowing the truth about risk, and a more balanced perspective.

The cost–benefit analysis and the cost-effectiveness analysis are also probability-based and, hence, there is a need to look beyond these analyses as well.

Basically, we may contrast two different approaches for reaching a good decision:

(i) Choose the alternative that maximizes/minimizes some specified criterion, for example E[NPV] or expected utility, reflecting the performance of the alternatives considered, values and uncertainties.

(ii) See decision-making as a process with formal decision and risk analyses providing decision support, followed by an informal managerial judgement and review process resulting in a decision.

The first approach is typical for the pioneers of the economic decision-making school (von Neumann and Morgenstern, 1944), and later the Bayesian decision-making theorists; see, for example Keeney and Raiffa (1993) and Lindley (1985). The second approach means that the analysis is strictly an aid for decision-making. This is comparable to the 'moderate' view, as mentioned by Fischhoff *et al.* (1995) and supported by a number of decision theorists; see, for example Watson and Buede (1987) and French and Rios Insua (2000). The decision-maker needs to place the results of the decision and risk analysis into a larger context of review and judgement. This does not mean that we cannot see examples where approach (i) may be appropriate, but considering varying degrees of the informal evaluation process preceding any decision, we may think of approach (i) as a special case of approach (ii).

References

Apostolakis, G.E. (1990) The concept of probability in safety assessments of technological systems. *Science*, **250**, 1359–1364.

Apostolakis, G.E. (2004) How useful is quantitative risk assessment? *Risk Analysis*, **24**, 515–520.

Aven, T. (2003) *Foundations of Risk Analysis*, John Wiley & Sons, Ltd, Chichester.

Aven, T. (2008) *Risk Analysis*, John Wiley & Sons, Ltd, Chichester.

Aven, T. (2009) Business loves risk assessments, Paper submitted for possible publication.

Aven, T. and Abrahamsen, E.B. (2007) On the use of cost-benefit analysis in ALARP processes. *International Journal of Performability Engineering*, **3**, 345–353.

Aven, T. and Flage, R. (2009) Use of decision criteria based on expected values to support decision-making in a production assurance and safety setting. *Reliability Engineering and System Safety*, **94**, 1491–1498.

Aven, T. and Vinnem, J.E. (2007) *Risk Management, with Applications from the Offshore Oil and Gas Industry*, Springer-Verlag, New York.

Bedford, T. and Cooke, R. (2001) *Probabilistic Risk Analysis*, Cambridge University Press, Cambridge.

Bohnenblust, H. (1998) Risk-based decision making in the transportation sector, in R.E. Jorissen and P.J.M. Stallen (eds), *Quantified Societal Risk and Policy Making*, Kluwer Academic Publishers, Dordrecht.

Cabinet Office (2002) Risk: Improving Government's Capability to Handle Risk and Uncertainty, Strategy unit report. UK.

Damodaran, A. (2002) *Investment Valuation*, 2nd edn, John Wiley & Sons, Inc., New York.

Environmental Assessment Institute (EAI) (2006) Risk and Uncertainty in Cost Benefit Analysis, Toolbox paper for the Environmental Assessment Institute, http://www.imv.dk/files/Filer/IMV/Publikationer/Rapporter/2006/risk_and_uncertainty.pdf (accessed April 2006).

Fischhoff, B. (1995) Risk perception and communication unplugged: twenty years of process. *Risk Analysis*, **15**, 137–145.

Fischhoff, B., Lichtenstein, S., Slovic, P. *et al.* (1981) *Acceptable Risk*, Cambridge University Press, New York.

French, S. and Ríos Insua, D. (2000) *Statistical Decision Theory*, Arnold, London.

Haimes, Y.Y. (2004) *Risk Modeling, Assessment, and Management*, 2nd edn, John Wiley & Sons, Inc., Hoboken, NJ.

HSE (Health and Safety Executive) (2001) *Reducing Risks, Protecting People – HSE's Decision-Making Process*, HSE Books, www.hse.gov.uk/dst/r2p2.pdf.

HSE (Health and Safety Executive) (2006) Offshore Installations (Safety Case) Regulations 2005 Regulation 12 – Demonstrating Compliance with the Relevant Statutory Provisions, Offshore Information Sheet No. 2/2006. www.hse.gov.uk/offshore/sheet22006.pdf.

Hull, J.C. (1980) *The Evaluation of Risk in Business Investment*, Pergamon Press, Oxford.

IRGC (International Risk Governance Council) (2005) White Paper on Risk Governance. Towards an Integrative Approach . Author: O. Renn with Annexes by P. Graham. International Risk Governance Council, Geneva.

ISO (2008) *Petroleum, Petrochemical and Natural Gas Industries: Production Assurance and Relibility Management*, ISO 20815:2008.

Jonkman, S.N., van Gelder, P. and Vrijling, J.K. (2003) An overview of quantitative risk measures for loss of life and economic damage. *Journal of Hazardous Materials*, **99**, 1–30.

Kaplan, S. (1991) Risk assessment and risk management – basic concepts and terminology, in R.A. Knief (ed.), *Risk Management: Expanding Horizons in Nuclear Power and Other Industries*, Hemisphere, New York, pp. 11–28.

Kaplan, S. and Garrick, B.J. (1981) On the quantitative definition of risk. *Risk Analysis*, **1**, 11–27.

Keeney, R.L. and Raiffa, H. (1993) *Decisions with Multiple Objectives: Preferences and Value Tradeoffs*, Cambridge University Press, Cambridge.

Levy, H. and Sarnat, M. (1994) *Capital Investment and Financial Decisions*, 5th edn, Prentice Hall, New York.

Lindley, D.V. (1985) *Making Decisions*, 2nd edn, John Wiley & Sons, Ltd, London.

Michaels, D. (2008) *Doubt is their Product*, Oxford University Press, New York.

O'Brien, M. (2000) *Making Better Environmental Decisions*, MIT Press, Cambridge, MA.

Paté-Cornell, M.E. (1996) Uncertainties in risk analysis: six levels of treatment. *Reliability Engineering and System Safety*, **54** (2–3), 95–111.

Reid, S.G. (1992) Acceptable risk, in D.I. Blockley (ed.), *Engineering Safety*, McGraw-Hill, New York, pp. 138–166.

Renn, O. (1998) Three decades of risk research: accomplishments and new challenges. *Journal of Risk Research*, **1** (1), 49–71.

Renn, O. (2008) *Risk Governance*, Earthscan, London.

Rosa, E.A. (1998) Metatheoretical foundations for post-normal risk. *Journal of Risk Research*, **1**, 15–44.

Shrader-Frechette, K.S. (1991) *Risk and Rationality: Philosophical Foundations for Populist Reforms*, University of California Press, Berkeley.

Stirling, A. (1998) Risk at a turning point? *Journal of Risk Research*, **1**, 97–109.

Stirling, A. (2007) Science, precaution and risk assessment: towards more measured and constructive policy debate. *European Molecular Biology Organisation Reports*, **8**, 309–315.

Tickner, J. and Kriebel, D. (2006) The role of science and precaution in environmental and public health policy, in E. Fisher, J. Jones and R. von Schomberg (eds), *Implementing the Precautionary Principle*, Edward Elgar Publishing, Northampton.

Treasury Board of Canada (1998) Benefit Cost Analysis Guide, Draft, www.tbs-sct.gc.ca/fin/sigs/Revolving_Funds/bcag/BCA2_e.asp (accessed July 1998).

Trigeorgis, L. and Mason, S.P. (2001) Valuing managerial flexibility, in E.S. Schwartz and L. Trigeorgis (eds), *Real Options and Investment Under Uncertainty: Classical Readings and Recent Contributions*, MIT Press, Cambridge, MA.

Varian, H.R. (1999) *Intermediate Microeconomics: A Modern Approach*, 5th edn, W.W. Norton, New York.

Vatn, J. (2007) Societal security – a case study related to a cash depot, in T. Aven and J. Vinnem (eds), *Risk, Reliability and Societal Safety (ESREL 2007)*, Vol. **3**, Taylor & Francis, London, pp. 2599–2605.

von Neumann, J. and Morgenstern, O. (1944) *Theory of Games and Economics*, Princeton University Press, Princeton, NJ.

Vose, D. (2008) *Risk Analysis: A Quantitative Guide*, 3rd edn, John Wiley & Sons, Ltd, Chichester.

Vrijling, J.K., van Hengel, W. and Houben, R.J. (1995) A framework for risk evaluation. *Journal of Hazardous Materials*, **43**, 245–261.

Walls, M.R. (2004) Combining decision analysis and portfolio management to improve project selection in the exploration and production firm. *Journal of Petroleum Science and Engineering*, **44**, 55–65.

Watson, S.R. and Buede, D.M. (1987) *Decision Synthesis – The Principles and Practice of Decision Analysis*, Cambridge University Press, Cambridge.

Further reading

Aven, T. (2009) Business loves risk assessments. Paper submitted for possible publication.

Aven, T. and Flage, R. (2009) Use of decision criteria based on expected values to support decision-making in a production assurance and safety setting. *Reliability Engineering and System Safety*, **94**, 1491–1498.

O'Brien, M. (2000) *Making Better Environmental Decisions*, The MIT Press, Cambridge.

19

The precautionary principle and risk management cannot be meaningfully integrated

There are many definitions of the precautionary principle. See, for example Löfstedt (2003) and Sandin (1999). The most commonly used definition is probably the 1992 Rio Declaration:

> In order to protect the environment, the precautionary approach shall be widely applied by States according to their capabilities. Where there are threats of serious or irreversible damage, lack of full scientific certainty shall not be used as a reason for postponing cost-effective measures to prevent environmental degradation.

Seeing beyond environmental protection, a definition as follows reflects what we believe is a typical way of understanding this principle:

> The precautionary principle is the ethical principle saying that if the consequences of an activity could be serious and are subject to scientific uncertainties, then precautionary measures should be taken or the activity should not be carried out at all.

The key aspect is that if there is a lack of scientific certainty about what will be the consequences of an activity, then actions are required, possibly abstention from the activity.

Misconceptions of Risk T. Aven

History of the precautionary principle

Precautionary 'thinking' has a long history (UN, 2005). The *Late Lessons from Early Warnings* report (Harremoës *et al.*, 2002) mentions the example of Dr John Snow who, in 1854, recommended removing the handle of a London water pump in order to stop a cholera epidemic. The evidence for the causal link between the spread of cholera and contact with the water pump was weak and not a 'proof beyond reasonable doubt'. This simple and relatively inexpensive measure, however, was very effective in halting the spread. The report mentions a series of other examples, such as asbestos (see below), where a precautionary approach could have saved many lives if early warnings of potential harm had been taken more seriously.

The precautionary principle, however, goes back to the 1970s. Both a Swedish and a German origin of the precautionary principle have been indicated. In Germany the precautionary principle ('Vorsorgeprinzip') is traced back to the first draft of a bill (1970) aimed at securing clean air. The law was passed in 1974 and covered all potential sources of air pollution, noise, vibrations and similar processes. The most unambiguous elaboration of the precautionary principle in German environmental policy is from a later date and reads: 'Responsibility towards future generations commands that the natural foundations of life are preserved and that irreversible types of damage, such as the decline of forests, must be avoided'. Thus: 'The principle of precaution commands that the damage done to the natural world (which surrounds us all) should be avoided in advance and in accordance with opportunity and possibility. *Vorsorge* further means the early detection of dangers to health and environment by comprehensive, synchronized (harmonized) research, in particular about cause and effect relationships..., it also means acting when conclusively ascertained understanding by science is not yet available. Precaution means to develop, in all sectors of the economy, technological processes that significantly reduce environmental burdens, especially those brought about by the introduction of harmful substances' (Bundesministerium des Innern, 1984).

Although the precautionary principle originated in Europe in the early 1970s, it was not until the 1992 United Nations Conference on Environment and Development that the principle received broad international recognition (Cameron, 2006). It was positioned as an underlying element of the broader framework of sustainable development. Since then, it has spread rapidly in multilateral agreements, international laws and domestic laws and policies dealing with: climate change, biodiversity, endangered species, fisheries management, wildlife, trade, food safety, pollution controls, chemicals regulation, exposure to toxins, and other environmental and public health issues (Peterson, 2006).

The example of asbestos

Nowadays it is known that asbestos is the main cause of mesothelioma, a disease with a very long incubation time which, once it manifests, is normally fatal within a year (*EEA 2001; UN 2005*). Health experts estimate that in the European Union (EU) alone, some 250 000–400 000 deaths from mesothelioma, lung cancer and asbestosis will occur over the next 35 years, as a consequence of exposure to asbestos in the past. Mining for asbestos began in 1879. At that time science was not aware of the dangers of asbestos. The annual production of asbestos worldwide grew to 2 million tonnes in 1998. Imports to the EU peaked in the mid 1970s and remained above 800 000 tonnes a year until 1980, falling to 100 000 tonnes in 1993. There is a delay of 50–60 years between the peak in import of asbestos and the peak in occurrence of mesothelioma in a country. Early warnings and actions are summarized in the following timeline:

- 1898. UK Factory Inspector Lucy Deane warns of harmful and 'evil' effects of asbestos dust.
- 1906. French factory report of 50 deaths in female asbestos textile workers and recommendation for controls.
- 1911. 'Reasonable grounds' for suspicion, from experiments on rats, that asbestos dust is harmful.
- 1911 and 1917. UK Factory Department finds insufficient evidence to justify further actions.
- 1930. UK 'Merewether Report' finds 66% of long-term workers in Rochdale factory with asbestosis.
- 1931. UK Asbestos Regulations specify dust control in manufacturing only and compensation for asbestosis, but this is poorly implemented.
- 1935–1949. Lung cancer cases reported in asbestos manufacturing workers.
- 1955. Research by Richard Doll (UK) establishes high lung cancer risk in Rochdale asbestos workers.
- 1959–1964. Mesothelioma cancer identified in workers, neighbourhood 'bystanders' and the public in South Africa, the UK and the USA, among others.
- 1998–1999. EU and France ban all forms of asbestos.
- 2000–2001. World Trade Organization upholds EU/French bans against Canadian appeal.

In the case of asbestos, a lack of full scientific proof of harm contributed to the long delay before action was taken and risk-reduction regulations were put in place. The early warnings of 1898–1906 were not followed up by any kind of precautionary action to reduce exposure to asbestos, nor by long-term medical and dust exposure surveys of workers that would have been possible at the time, and which would have helped strengthen the case for tighter controls on dust levels. A Dutch study has estimated that a ban in 1965, when the mesothelioma hypothesis was plausible but unproven, instead of in 1993 when the hazard of asbestos was widely acknowledged, would have saved the country some 34 000 victims and €19 billion in building costs (clean-up) and compensation costs. This is in a context of 52 600 victims and €30 billion in costs projected by the Dutch Ministry of Health over the period 1969–2030. Today, a substantial legacy of health and contamination costs has been left for both mining and user countries, while asbestos use continues now largely in developing countries.

Different interpretations of the precautionary principle

The precautionary principle may be interpreted in different ways (see Cameron, 2006):

1. The weak versions allow for consideration of the costs of the precautionary measures. A balance has to be struck between benefits and costs. The requirement to justify the need for action (the burden of proof) generally falls on those advocating precautionary action. No mention is made of assignment of liability for the possible harm.
2. Strong versions justify or require precautionary measures and some also establish liability for the possible harm, which is effectively a strong form of 'polluter pays'. For example, the Earth Charter (2000) states: 'When knowledge is limited apply a precautionary approach .... Place the burden of proof on those who argue that a proposed activity will not cause significant harm, and make the responsible parties liable for environmental harm.' Reversal of proof requires those proposing an activity to prove that the product, process or technology is sufficiently 'safe' before approval is granted. Requiring proof of 'no harm' before any action proceeds implies that the public is not prepared to accept any risk, no matter what economic or social benefits may arise (Peterson, 2006). At the extreme, such a requirement could involve bans and prohibitions on entire classes of potentially threatening activities or substances (Cooney, 2005).

Over time, there has been a gradual transformation of the precautionary principle from what appears in the Rio Declaration to a stronger form that arguably acts as restraint on development in the absence of firm evidence that it will do no harm (Cameron, 2006).

Implementation of the precautionary principle, in particular the strong versions, could result in increasing costs and delayed innovation and negatively affect the viability of innovative industries and those that depend on their products. This may encourage these industries to change their activities or relocate to other jurisdictions with less stringent standards of proof, resulting in a loss in capability in the home country (Cameron, 2006).

There is no consensus about the interpretation of the precautionary principle, and its meaning and use are strongly debated; see, for example, Fisher *et al.* (2006), Sandin (1999), Sandin *et al.* (2002), Löfstedt (2003), Sundstein (2005), Wiener and Rogers (2002), Hammit *et al.* (2005) and Wilson *et al.* (2006). Sandin (1999) lists more than 20 definitions adopted. It is, however, possible to identify four dimensions of the principle that apply to all of these definitions (Sandin, 1999): (i) the threat dimension, (ii) the uncertainty dimension, (iii) the action dimension and (iv) the command dimension. But the phrases expressing these dimensions vary in precision and strength. In the present review and analysis we are particularly concerned with the uncertainty dimension. What does uncertainty mean? Sandin (1999) refers to a threat which is uncertain, but all threats are uncertain: we do not know when the next event will occur and what will be the consequences of the event. The key concept is scientific uncertainties, but what does this term mean?

Scientific uncertainties

As the focus is on the future consequences of the activity, there would be no (or at least very few) cases with known outcomes. Hence, scientific certainty must mean something else – and three natural candidates are (Aven, 2006):

(i) knowing which *type of* consequences could occur;

(ii) being able to predict the consequences with sufficient accuracy;

(iii) having accurate descriptions or estimates of the real risks, interpreting the real risks as the consequences of the activity.

Adopting one of these interpretations, the precautionary principle could be applied either when we do not know the type of consequences that could occur, or we have poor predictions of the consequences, risk descriptions or estimates. As an example, let us think of the issue about starting year-round petroleum activities in the Lofoten/Barents Sea area off Northern Norway. The Norwegian government is considering whether year-round activities should be allowed in specific sectors of the area, which are vulnerable areas ecologically. Then following (i) and using broad categories of consequences, we cannot apply the precautionary principle as we know the type of consequences of this activity. As a result of the activity, some people could be killed, some injured, an oil spill could occur causing damage to the environment, and so on. Different categories of this damage could be defined. Hence, by grouping of categories and types of consequences the possible lack of scientific certainty is 'eliminated'.

However, in this case, many biologists would say that there is some lack of knowledge on what will be the consequences for the environment, given an oil spill. This lack of scientific certainty could be classified as rather small, but that would be a value statement and people and parties could judge this differently.

The point is that there is some scientific uncertainty about what will be the consequences of an oil spill. Consider the consequences of an oil spill on fish species, and let Z denote the recovery time for the population of concern, with Z being infinity if the population does not recover. Then there is scientific certainty according to criterion (ii) if there is scientific consensus about a function (model) f such that $Z = f(X_1, X_2, \ldots)$ with high confidence, where X_1, $X_2, \ldots$ are some underlying factors influencing Z. Such factors could relate to the possible occurrence of a blowout, the amount and distribution of the oil spilled on the sea surface, the mechanisms of dispersion and degradation of oil components, and the exposure and effect on the fish species. For selected values of the Xs, we can use f to predict the consequences Z. The precautionary principle applies when it is difficult to establish such a function f – the scientific discipline does not sufficient have knowledge to obtain 'scientific certainty' on how the high-level performance, in this case measured by Z, is influenced by the underlying factors. Models may exist, but they are not broadly accepted in the scientific community.

Scientific consensus in this sense does not mean that the consequences (Z) can be predicted with accuracy, when not conditioned on the Xs. Unconditionally, the consequences (Z) are uncertain, and this uncertainty is defined by the uncertainties of the factors X.

In practice, there would always be some degree of lack of scientific certainty. Hence, the question of valuing this degree arises. How important is the lack of scientific certainty? How accurate does the model f need to be? How can we measure the accuracy?

There are no clear answers to these questions. Different people and parties would judge these issues differently. There are no sharp limits stating that a specific level is not acceptable and that the precautionary principle should apply.

Hence, referring to the precautionary principle implies a judgement, expressing that we find the lack of scientific uncertainty, that is, the lack of knowledge related to how the consequences of the activity are influenced by the underlying factors, to be so significant that precautionary measures are required or the activity should not be carried out. The risk analysis results, producing predictions and uncertainty assessments, provide input to such a judgement.

Lay people's risk perception may influence the decision-maker and his/her attitude to the importance of various aspects of the risk picture. This applies in particular to the weight put on the lack of understanding of how the consequences of the activity are influenced by the underlying factors. Hence, lay people's risk perception may also affect the application of the precautionary principle.

To study criterion (iii), suppose that p represents the 'real' risk, quantified by the probability distribution of Z, and let p^* be an estimate of p derived by a detailed risk analysis of the activity. Since the uncertainties in this estimate are considered large, relative to the true p, the precautionary principle may be applied following criterion (iii).

This criterion means that the precautionary principle would be applied in most practical cases as the probability estimates are subject to large uncertainties. Let us return to the blowout example. We may have a strong scientific knowledge about the phenomena causing environmental damage in case of an oil spill, but large uncertainties about the blowout probability. Is the precautionary principle still applicable? Yes, according to criterion (iii), but not according to criterion (ii).

Except for situations where it is possible to sample a large number of similar items, the probability estimate would be subject to large uncertainties. Thus, there is a lack of scientific certainty and we may apply the precautionary principle. For the Barents Sea example, the risk estimates would be subject to relatively large uncertainties, and the precautionary principle would, therefore, be applicable.

This criterion is based on a traditional statistical perspective on probability and we have, in earlier chapters, discussed the problem of defining and interpreting the underlying risk and probabilities. In the Barents Sea example, the probability p expresses the proportion of 'experiments' in which the recovery time Z exceeds a specific number. The population is a fictional population generated by a thought-constructed experiment in which we simulate the activity in the Barents Sea over and over again, with some aspects being stochastic and some other aspects considered a part of the frame conditions of the experiment. For example, the performance of the workers offshore may vary, but the working positions are considered constant. If we lack accurate estimates of this underlying thought-constructed probability, we may apply the precautionary principle. As already noted, this means that for most complex situations in practice we may apply the precautionary principle, if this perspective is adopted, since the estimate would be subject to large uncertainties.

If knowledge-based probabilities are used, we may reformulate criterion (iii) as having expert consensus about the probabilities. However, the conclusion would often be the same as in the relative frequency case, as different experts produce quite different probability assignments.

Even if we restrict attention to the probability of a blowout, criterion (iii) is likely to lead to the application of the precautionary principle as there would be relatively large uncertainties about the probability estimates (different probability assignments). Nonetheless, the phenomena leading to blowout are well understood. Obviously, it is difficult to predict when a blowout is to occur, but should we refer to this uncertainty as scientific uncertainty? Some would perhaps refer to this uncertainty as stochastic (aleatory), but a closer look reveals that it is not random and irreducible. It is certainly possible to influence the technical, human and organizational factors that cause a blowout to occur. There are risks and uncertainties associated with a possible occurrence of a blowout, but should the precautionary principle be applicable for all types of uncertainties and risk?

Cautionary principle

Is the precautionary principle just a term used to capture aphorisms such as 'an ounce of prevention is worth a pound of cure', 'better safe than sorry', and 'look before you leap' (Science and Environmental Health Network, 2000)? No, these aphorisms are more in line with the cautionary principle (HSE, 2001; ILGRA,

2002; Aven, 2006) stating that, in the face of risks and uncertainties, caution should be the ruling principle. This principle is implemented in all industries through regulations and requirements. For example, in the Norwegian petroleum industry it is a regulatory requirement that the living quarters on an installation should be protected by fireproof panels of a certain quality, for walls facing process and drilling areas. This is a standard adopted to obtain a minimum safety level. It is based on established practice of many years of operation of process plants. A fire may occur – it represents a hazard for the personnel, and in the case of such an event, the personnel in the living quarters should be protected. The assigned probability for the living quarters on a specific installation being exposed to fire may be judged as low, but we know that fires occur from time to time in such plants. The justification is experience from similar plants and sound judgement. A fire may occur, since it is not an unlikely event, and we should be prepared. We need no references to cost–benefit analysis. The requirement is based on cautionary thinking.

Risk analyses, cost–benefit analyses and similar types of analyses are tools providing insights into risks and the trade-offs involved. But they are just tools – with strong limitations. Their results are conditioned on a number of assumptions and suppositions. The analyses do not express objective results. Being cautious also means reflecting this fact. We should not put more emphasis on the predictions and assessments of the analyses than can be justified by the methods being used.

In the face of uncertainties related to the possible occurrences of hazards and threats, we are cautious and adopt principles of safety management, such as

- robust design solutions, such that deviations from normal conditions do not lead to hazardous situations and accidents,
- design for flexibility, meaning that it is possible to utilize a new situation and adapt to changes in the frame conditions,
- implementation of safety barriers, to reduce the negative consequences of hazardous situations if they should occur, for example a fire,
- improvement of the performance of barriers by using redundancy, maintenance/testing, and so on,
- quality control/quality assurance,
- the precautionary principle, by which in the case of lack of scientific certainty about the possible consequences of an activity, measures should be taken or we should not carry out the activity,
- the ALARP principle, which says that the risk should be reduced to a level which is as low as reasonably practicable.

The level of caution adopted will, of course, have to be balanced against other concerns such as costs. However, all industries would introduce some minimum requirements to protect people and the environment, and these requirements can be considered justified by reference to the cautionary principle.

These are measures in a safety context. Other types of measures are required in other types of applications, for example investment analysis and finance. Examples include diversification of investments to reduce risk and uncertainties, and maximum allowable investments.

Adopting this terminology, the precautionary principle is considered a special case of the cautionary principle, as it is applicable in cases of *scientific uncertainties* about the possible consequences of the activity being considered (Aven, 2006). The distinction between the cautionary principle and the precautionary principle is not common, but is considered useful for separating the attitude and actions in case of risks and uncertainties, from the attitude and actions in the special case of scientific uncertainties. Many researchers as well as lay people seem to use the term precautionary principle for both cases.

The cautionary and precautionary principles' place in risk management

Among most economists and decision analysts, the theoretical framework for obtaining good decisions is the expected utility theory, based on the use of subjective probabilities. Attention should be on $Eu(X)$, where u is the utility function and X is the outcome. In this framework there is no place for the application of the precautionary principle, as the expected utility is the appropriate guidance for the decision-maker. Uncertainties and the weights put on these uncertainties are properly taken into account using this theory.

However, this is a theory, and it is difficult to apply in practice. People do not behave according to this theory. This is well known, and different alternative frameworks have been suggested. Many economists would refer to cost–benefit analysis as the appropriate practical tool to guide decision-makers. By transforming all values to monetary values and calculating expected net present values, a consistent procedure is obtained for making decisions, which are believed to be good decisions.

Again, in this framework there is no place for the application of the precautionary principle, as cost–benefit analysis is the appropriate guidance for the decision-maker. However, few people would conclude that the cost–benefit analyses and related tools provide clear answers. They have limitations and are based on a number of assumptions and presumptions, and their use is based not only on scientific knowledge, but also on value judgements involving ethical, strategic and political concerns. The analyses provide support for decision-making, leaving the decision-makers to apply decision processes outside the direct applications of the analyses. It is necessary to look beyond the expected values. We refer to the previous chapter.

Using risk acceptance criteria $P < p_0$, as was discussed in the previous chapter, a project or arrangement is accepted provided that the estimated probability is below a certain threshold. Again there is no reference to the precautionary principle. As for expected net present value, it is possible to adjust the criteria to better reflect uncertainties but, as argued in the previous chapter, uncertainty needs to be considered beyond the probabilities and expected values. The issue is how the uncertainties should be taken into account in the decision-making process. The precautionary principle is a way of dealing with the uncertainties. Remember the four dimensions of the principle (Sandin, 1999): (i) the threat dimension, (ii) the uncertainty dimension, (iii) the action dimension and (iv) the command dimension. The uncertainties are often restricted to scientific uncertainties, but we also see reference to uncertainties about threats and consequences in general. A main goal of risk management is to balance uncertainty with other concerns and make good decisions under uncertainty. Hence, the cautionary and precautionary principles can be viewed as principles of risk management, which say what to do in case of uncertainties. But mechanical procedures for what to do in case of uncertainties should not be implemented, as discussed in the previous chapter. It is a management task (management review and judgement) to give the proper weight to the uncertainties, and this weight has to be seen in a broad context where all concerns are reflected. This also means considerations beyond the probability analysis.

Cash depot example (continued from previous chapter)

In this case, considerable weight given to the cautionary principle would have led to removal of the NOKAS facility. There is significant uncertainty associated with whether we will experience an attack linked to the NOKAS facility in the future, and what the course of events will be if an attack should actually occur. This fact calls for the application of the cautionary principle, but the level of caution adopted is balanced against other concerns, such as costs. There is no reference to the precautionary principle, as we are not dealing with scientific uncertainties. It is possible (in theory) to specify with high confidence a function such that the consequences can be predicted with high accuracy. □

As discussed above, it is common to restrict the precautionary principle to the lack of understanding of how the consequences of the activity are influenced by the underlying factors (Aven, 2006). Any reference to being able to accurately estimate probabilities should be avoided, as that leads to a meaningless discussion of accuracy in probability estimates. We have to acknowledge that it is not possible to establish science-based criteria for when the precautionary principle should apply. Judging when there is a lack of scientific certainty is a value judgement. In the face of uncertainty, analysts and scientists need to do a good job to express the uncertainties, enabling the decision-maker to obtain an informative basis for his or her decision. There is a large potential for improvement on risk and uncertainty descriptions and communications. Many analysts and scientists have severe problems in dealing with uncertainties, as do many statisticians. We see being aware of the different perspectives on risk, and using these in the descriptions and communications, as a key element in improving the present situation.

Restricting the precautionary principle to scientific uncertainties, a strong implementation version is the natural interpretation. Refer to the introduction to this chapter: strong versions justify or require precautionary measures and some also establish liability for the possible harm, which is effectively a strong form of 'polluter pays'. These versions mean to place the burden of proof on those who argue that a proposed activity will not cause significant harm, and make the responsible parties liable for possible harm.

But a management review and judgement is required as the definition and the importance of the scientific uncertainties always need to be weighed against other concerns. Obviously, a strict adherence to such a strong implementation procedure of the precautionary principle would lead to no development or growth. A society requires risk appetite, as discussed in Chapter 17. The challenge is to find the proper balance, and such a balance cannot be established by strict mechanical procedures.

Summary

The precautionary principle needs to be considered in the context of risk management, which provides a framework and approach for setting the best course of action under uncertainty. Such an approach could assist in determining when and how the principle should be applied to manage risk and uncertainty and balance different concerns such as costs and safety. This could enable the greatest benefits and returns to be achieved. Development and innovation should be supported, but would suffer from too strict an implementation of the precautionary principle.

Cautionary measures are required in the case of risks and uncertainty. We implement safety barriers and emergency preparedness in case of a hydrocarbon leakage in a process plant, but the process and phenomena are well understood and the precautionary principle is not applicable. There are no scientific uncertainties. The cautionary principle applies in the case of risk and uncertainty, and the precautionary principle in the case of scientific uncertainties.

References

Aven, T. (2006) On the precautionary principle, in the context of different perspectives on risk. *Risk Management, An International Journal*, **8**, 192–205.

Bundesministerium des Innern (1984) Dritter Immissionsschutzbericht, Drucksache Bonn 10/1345.

Cameron, L. (2006) Environmental risk management in New Zealand – Is there scope to apply a more generic framework? New Zealand Treasury Policy Perspectives Paper 06/06, July.

Cooney, R. (2005) From promise to practicalities: the precautionary principle on biodiversity conservation and sustainable use, in R. Cooney and B. Dickson (eds), *Biodiversity and the Precautionary Principle: Risk and Uncertainty in Conservation and Sustainable Use*, Chapter 1, Earthscan, London.

Earth Charter (2000) The Earth Charter Initiative: Values and Principles for a Sustainable Future, http://www.earthcharterinaction.org/2000/10/the_earth_charter.html (accessed September 2009).

EEA (European Environment Agency) (2001) Late Lessons from Early Warnings: The Precautionary Principle 1896–2000. Environmental Issue Report No. 22.

Fisher, E., Jones, J. and von Schomberg, R. (eds) (2006) *Implementing the Precautionary Principle Perspectives and Prospects*, Edward Elgar, Cheltenham and Northampton.

Hammitt, J.K., Wiener, J.B., Swedlow, B. *et al.* (2005) Precautionary regulation in Europe and the United States: a quantitative comparison. *Risk Analysis*, **25**, 1215–1228.

Harremoës, P., Gee, D., MacGarvin, M. *et al.* (eds) (2002) *The Precautionary Principle in the 20th Century: Late Lessons from Early Warnings*, Earthscan, London.

HSE (2001) *Reducing Risk, Protecting People*, HSE Books.

ILGRA (2002) The Precautionary Principle: Policy and Application, Inter-Departmental Liaison Group on Risk Assessment.

Löfstedt, R.E. (2003) The precautionary principle: risk, regulation and politics. *Transactions of IchemE*, **81**, 36–43.

Peterson, D. (2006) Precaution: principles and practice in Australian environmental and natural resource management. Productivity Commission. Presidential Address, 50th Annual Australian Agricultural and Resource Economic Society Conference, Manly, New South Wales.

Sandin, P. (1999) Dimensions of the precautionary principle. *Human and Ecological Risk Assessment*, **5**, 889–907.

Sandin, P., Peterson, M., Hansson, S.O. *et al.* (2002) Five charges against the precautionary principle. *Journal of Risk Research*, **5**, 287–299.

Science and Environmental Health Network (2000) The Precautionary Principle: A Common Sense Way to Protect Public Health and the Environment, January.

Sundstein, C.R. (2005) *Laws of Fear. Beyond the Precautionary Principle*, Cambridge University Press, Cambridge.

UN (2005) The Precautionary Principle, United Nations Educational, Scientific and Cultural Organization, UNESCO, World Commission on the Ethics of Scientific Knowledge and Technology (COMEST).

Wiener, J.B. and Rogers, M.D. (2002) Comparing precaution in the United States and Europe. *Journal of Risk Research*, **5**, 317–349.

Wilson, K., Leonard, B., Wright, R. *et al.* (2006) Application of the precautionary principle by senior policy officials: results of a Canadian survey. *Risk Analysis*, **26**, 981–988.

Further reading

Aven, T. (2009) On different types of uncertainties in the context of the precautionary principle, Paper submitted for possible publication.

20

Conclusions

The previous chapters have discussed many conceptions and misconceptions about risk, and the reader may already have a good picture of this book's main findings and conclusions. In the following, a summary of these findings and conclusions is presented.

How risk is defined

Uncertainty is the main component of the risk concept, together with events and associated consequences. Different definitions of risk can support such a perspective, as discussed in Aven and Renn (2009a). One candidate, call it definition I, is (Aven, 2007):

By risk we understand the two-dimensional combination of

(i) events A and the consequences of these events C, and

(ii) the associated uncertainties U (whether A will occur and what value C will take).

We refer to this as the (A, C, U) perspective.

We may rephrase this definition by saying (call it definition I′) that risk associated with an activity is to be understood as (Aven and Renn, 2009a) *uncertainty about and severity of the consequences of an activity*, where severity refers to intensity, size, extension, scope and other potential measures of magnitude, and is with respect to something that humans value (lives, the environment, money, etc.). Losses and gains, for example expressed by money or the number of fatalities, are ways of defining the severity of consequences. It is important to note that the uncertainties relate to the events and consequences – the severity is just a way of characterizing the consequences. Something that humans value is at stake – undesirable events and consequences could occur. However, the activity

Misconceptions of Risk T. Aven

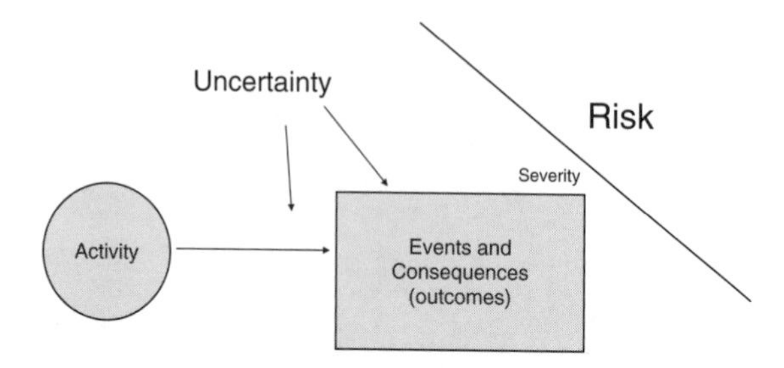

Figure 20.1 Illustration of the risk definition (I, I′) (Aven and Renn, 2009a).

may also lead to desirable outcomes. The main features of this risk perspective are illustrated in Figure 20.1.

These definitions have the following properties:

- They accommodate both undesirable and desirable outcomes.
- They address uncertainties instead of probabilities and expected values.
- They are not restricted to specific consequences and quantities.

We consider these properties essential for the risk concept to provide a foundation for risk research and risk management. The definitions acknowledge the challenge of establishing a precise measure of uncertainty as well as a precise measure of the stakes involved (Rosa, 1998). The definitions do not fall into the extreme of total subjectivism and relativism, and they are pretending that risk is a measurable object similar to other physical entities (Aven and Renn, 2009a).

According to this perspective, risk does not exist independently of the assessor, as the uncertainties are somebody's uncertainties. The fact that there are uncertainties about the events and consequences may be considered a state of the world, but any judgement or description of these uncertainties would depend on the assessor. Consider the lung cancer example in Chapter 5. Would cancer exist if the conditions of cancer – the uncontrolled growth of cells – had not been detected (we may call this state of the person 'undetected cancer')? Yes, cancer exists, but this is not risk according to our definition. Risk, according to our definition, adds the uncertainty (knowledge) dimension. This uncertainty can be based on observations and/or causal knowledge about dose and effect.

The fact that both uncertainty and severity imply a reference to the assessor should not be confused with the postmodern notion that risk knowledge is socially constructed. Clearly, there are events (dangers) out there even if they have not yet been perceived (e.g. new, undiscovered diseases). Rosa (1998) writes:

> that no dangers exist beyond those we perceive or beyond our current measurement sophistication is a difficult argument to sustain. It is

> another version of the 'argument from ignorance' fallacy (Shrader-Frechette, 1985), meaning that if we are ignorant of some danger there is little basis for claiming the danger exists at all.

The assumption that threats are only real if we perceive them as risks is counter-intuitive and easy to refute. Yet, does our definition of risk fall prey to this fallacy when it emphasizes the subjective nature of both uncertainty and severity? This is not the case. Our risk concept does not prescribe the consequences. There could be large uncertainties related to specific consequences or events that may or may never materialize. The uncertainties may be a result of 'known uncertainties' – we know what we do not know – and 'unknown uncertainties' (ignorance or non-knowledge) – we do not know what we do not know. For a specific activity an undiscovered disease may occur. This risk exists for the assessor provided that the assessor opens up to the possible occurrence of such consequences.

It is common to distinguish between (i) the state of the world (ontology), (ii) the knowledge about that state (epistemology) and (iii) normative judgement (what ought to be done). Our definition covers (i) and (ii), but leaves (iii) to the risk treatment. There is no normative aspect involved in the merging of uncertainty and severity. We do not propose a specific rule of combining both components, such as multiplication. This would have implied a value judgement, for example that the components should be assigned weights according to an expected value formula.

We may compare our definition with a probability distribution of a quantity X (e.g. expressing the number of fatalities) based on knowledge-based (subjective) probabilities. This probability distribution exists for the assessor, and covers both dimensions (i) and (ii). However, our definition is more general as probability is replaced by uncertainties and the consequences are not restricted to a set of defined random quantities.

How risk is described

Risk is described by (A, C, U, P, K), that is, by events and consequences, associated uncertainties (whether A will occur and what value C will take), knowledge-based probabilities with reference to a standard, and K the background knowledge that U and P are based on. The probabilities are a tool to express uncertainties but there is a need to look beyond the probabilities and associated expected values when assessing uncertainties. Uncertainty factors may be 'hidden' in K. In addition, sensitivities (S) are included to show how the results depend on variation in input assumptions and conditions (parameters). To reflect this, the risk description is adjusted to (A, C, U, P, S, K) (Flage and Aven, 2009). Further details are given when we discuss the framework for risk assessment below (see also Chapter 12).

The role of risk assessments and cost–benefit analysis in risk management

Risk acceptability and tolerability cannot be defined based on risk assessments alone. A balance has to be struck between different concerns, including costs. In addition, risk is more than calculated numbers. Consequently, it is impossible to restrict the risk evaluation to simple comparisons between numbers. Uncertainties beyond the probabilities need to be taken into account.

Cost–benefit analysis and cost-effectiveness analysis can provide useful decision support, but it is necessary to acknowledge that the analyses do not adequately reflect risk and uncertainties. Assessments of risk and uncertainties beyond these analyses are required.

Overall risk acceptance criteria are not used. Holistic considerations of arrangements and measures are conducted on the basis of analyses and all relevant concerns are taken into account. The constraints and limitations of the analyses constitute a part of the decision basis. This evaluation is referred to as the *management review and judegment*; see Chapter 18. These principles are illustrated in Figure 20.2.

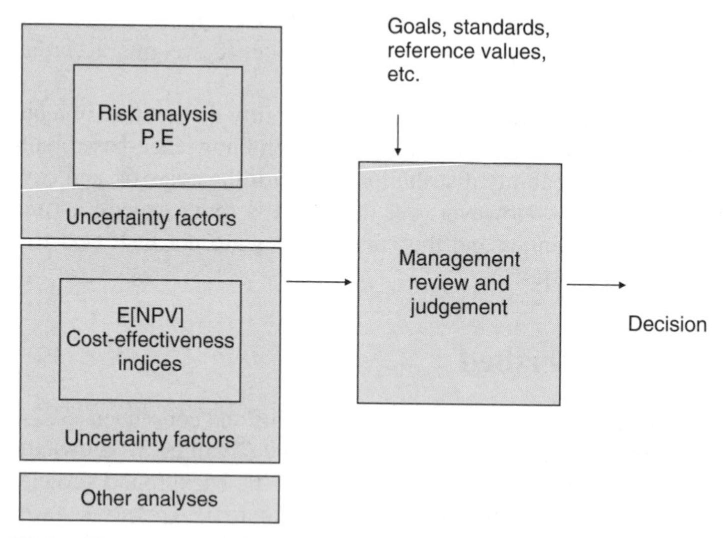

Figure 20.2 Illustration of the main principle of the risk management approach: P: Knowledge-based probability with reference to a standard, E: expectation.

Below we present a framework for risk assessments which is based on the above definitions and principles and summarized in Aven (2009). The framework represents a new stage development of the overall scientific structure of risk assessments. The 'best estimate approach' and the 'probability of frequency approach' are two common existing stages. The best estimate approach restricts attention to point estimates of frequentist probabilities and expected values. The probability of frequency approach extends the analysis by describing epistemic uncertainties about the underlying frequentist probabilities and expected values, using subjective (knowledge-based) probabilities (Kaplan and Garrick, 1981); see Chapter 18. The approach may be considered a reconciliation of the relative frequency based and the Bayesian schools of thought, which have clashed for many years over the definition of probability, sowing considerable confusion over the definition of risk and the limits of probabilistic risk analysis (Paté-Cornell, 1996). Bayesian theorists would not refer to the 'frequentist probabilities' as probabilities, but as chances or propensities (Singpurwalla, 2006). However, from a practical point of view, an analyst would not see much difference between the Bayesian theorist view and the probability of frequency approach. These two stages correspond to level 3 and level 5 in the uncertainty treatment classification system discussed by Paté-Cornell (1996).

The new stage is characterized by a shift in thinking from probability/chance to uncertainty as the starting point for the analysis.

Framework for risk assessment

Figure 20.3 provides a more detailed description of the risk assessment framework. The key quantities of interest are denoted Z (which could be a vector). To assess Z, a model $G(X)$ is introduced which links a set of input quantities X to Z. To describe the uncertainties, knowledge-based (subjective) probabilities are used. Using the model G, an uncertainty description is obtained for Z. The tool used for this purpose could be an analytical approach or Monte Carlo simulation.

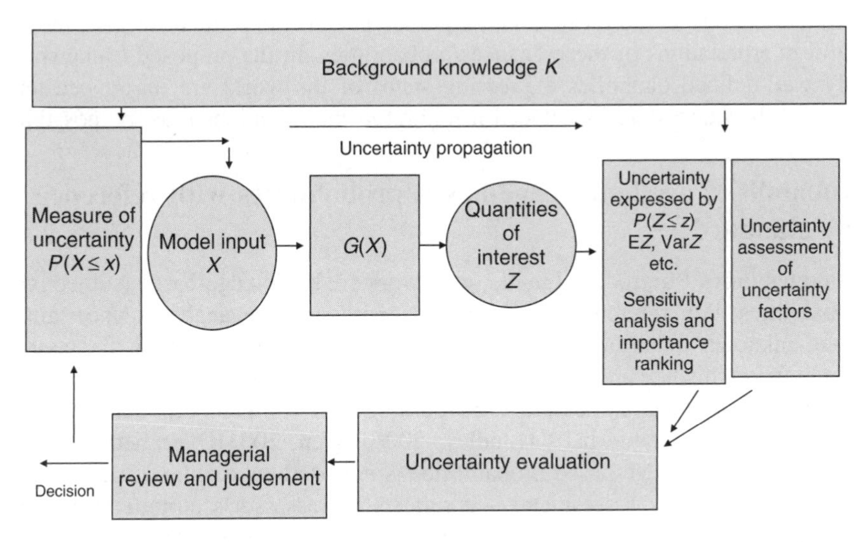

Figure 20.3 Structure of the risk assessment framework (Aven, 2009).

Many features of the framework have been discussed in previous chapters; for example, Chapter 16 studies sensitivity analysis and importance ranking. In the following we will address these features of the framework:

- The quantities X and Z are well defined.
- All probabilities are knowledge-based (subjective) probabilities with reference to a standard.
- Assessments of uncertainty factors.
- Model uncertainty (inaccuracy) is not quantified.
- The need for a managerial review and judgement.

The quantities X and Z are well defined

The quantities X and Z express states of the 'world', that is, quantities of a physical reality or nature, that are unknown at the time of the analysis but will, if the system being analysed is actually implemented, acquire some value in the future, and possibly become known. X and Z must have some true, objective values. No ambiguity can be present. In our view, uncertainty assessment of quantities for which true and precisely defined values do not exist cannot be a basis for a scientific risk assessment. This is a key assumption of the framework, and supported by, for example Bedford and Cooke (2001).

If chances (probabilities with a relative frequency interpretation) are introduced they must be considered unknown properties of the world and be treated as X and Z in the framework. There is a tendency in many risk assessment applications to introduce chances and frequentist probabilities also when such constructions are based on fictional populations and it is difficult to give meaningful interpretations of these chances/probabilities. In the proposed framework, only well-defined quantities expressing states of the world are introduced and assessed. If chances are introduced meaningful interpretations must be possible.

Probabilities are knowledge-based probabilities with reference to a standard

All probabilities P introduced in the framework are knowledge-based (subjective) probabilities with reference to a standard expressing the analyst's uncertainty about unknown quantities X and Z. Following this interpretation, the assessor compares his/her uncertainty about the occurrence of the event A with the standard of drawing at random a favourable ball from an urn that contains $P(A) \cdot 100\%$ favourable balls (Lindley, 2006; Aven, 2003). The betting interpretation of knowledge-based probabilities is not used, as it extends beyond the realm of uncertainty assessments – it reflects the assessor's attitude to money and the gambling situation which means that analysis (evidence) is mixed with values; see Chapter 12.

Frequentist probabilities P_f do not exist in this framework, but proportions of infinite or very large populations of similar (technically, exchangeable) units to those considered exist. Such proportions we refer to as chances, in line with Bayesian terminology (Singpurwalla, 2006). They are treated as unknown quantities X and Z.

The framework probabilities $P(Z \leq z)$ and so on cannot be frequentist probabilities P_f as such probabilities are, in fact, not somebody's measure of uncertainty, but a way of expressing variation within a real or thought-constructed infinite (or very large) population of similar units to those studied; see Chapter 12.

The probabilities $P(Z \leq z)$ and so on express epistemic uncertainties. The variation in the populations of similar units to the one studied, that, for example, generates the true value of P_f, is often referred to as aleatory (stochastic) uncertainty. This uncertainty is, however, not an uncertainty for the analysts and in the following we will refer to it as a variation or a chance.

Assessments of uncertainty factors

Different probability-based measures are used to describe risk, such as expected values, variance and quantiles. But a full risk description needs to see beyond these P measures. All probabilities are conditional on a background knowledge K, which includes assumptions and suppositions, and in particular the model G. This background knowledge is an integral part of the results of the analysis and all probabilities need to be considered in relation to K. The framework requires a separate identification and assessment of potential uncertainty factors hidden in K; see Chapter 12.

Model uncertainty is not quantified

A model is a representation of the world. It is introduced to obtain insights about the phenomena being studied and to quantify the uncertainties. The model should describe the world sufficiently accurately, but also simplify complicated features and conditions. There is always a balance to be made between these concerns. In the framework, the model G is introduced linking X and Z. The difference between $G(X)$ and Z is the 'error' introduced by G. We may refer to this 'error' as model inaccuracy or model uncertainty. It obviously needs to be addressed as the uncertainty assessments are conditional on the use of this model.

When testing the model we will focus on this error and if observations of Z are available, we will compare the assessments of Z, which are conditional on the use of the model G, with these observations. The result of such a comparison provides a basis for improving the model and accepting it for use. But at a certain stage we accept the model and use it to support decision-making concerning risk acceptance and the choice of arrangements and measures. Then it has no meaning in quantifying model uncertainty (inaccuracy). If the model is not considered good enough for its purpose, it should be improved. Instead of specifying $P(Z \leq z)$ directly, we compute $P(G(X) \leq z|K)$ and G is a part of the background knowledge K. Key assumptions underpinning the model can be identified as uncertainty factors, and highlighted as a part of the risk and uncertainty picture, but the importance of these factors are not quantified. The performance of a model must always be seen in light of the purpose of the analysis. A crude model can be preferred to a more accurate model in some situations if the model is simpler and it is able to identify the essential features of the system performance.

However, model uncertainty quantification in the sense of model weighting (refer to the example in Chapter 13) is covered by the framework. Model weighing is a completely different issue than quantification of model inaccuracy. When using the framework to compute $P(Z \leq z)$ and so on we accept the use of specific models and possible procedures for weighting the models. The models and procedures are part of the background knowledge K.

The need for an uncertainty evaluation and managerial review and judgement

In the uncertainty evaluation, a broad uncertainty description is provided, covering probabilities and related background knowledge, the results from sensitivity analyses, importance ranking, as well as the assessment results for the uncertainty factors. The evaluation provides input to a broader managerial review and judgement, which concludes on the implications of the analysis and balances different concerns. The result is, for example an acceptance of the risk and uncertainties related to an activity, the need for design changes, the choice of an alternative, and so on.

It is common in risk management to compare the results of the risk assessments with relevant decision criteria, such as requirements of the form $P < p_0$, where p_0 is a fixed number. These requirements could, for example, express the requirement that the risk level of an activity should not exceed a specified level.

Now, given that P is a knowledge-based probability, can we justify basing our decision on a direct comparison of the form $P < p_0$? No, there is a need for a process that extends beyond the probability analysis. The probabilities are dependent on the background knowledge, the assumptions and suppositions made, including the model G. There is a need for a broader process which sees the results of the assessment in a larger context, taking into account the limitations of the model, the difficulties in specifying probabilities for some quantities, and so on. This management review and judgement is a process extending beyond the domain of the uncertainty analysis. The sensitivity analyses constitute an important input to such a broad review and judgement process.

Industry practices (see de Rocquigny *et al*., 2008) also acknowledge the need for broader evaluation processes, in particular for situations characterized by large uncertainties. If the focus is on situations in which there is strong modelling expertise, knowledge and/or data it is easier to justify the use of quantitative analysis linked to decision criteria of the form $P < p_0$. However, even in cases with strong modelling expertize and much data, it is essential to avoid mechanical procedures for decision-making based on probabilistic (or other) criteria in isolation. The probabilities P are not objective values, and there could be other concerns for the decision-maker than risk and uncertainties.

The cautionary and precautionary principles have important roles to play in risk management, to give the proper weight to uncertainties. We implement cautionary measures in the face of risk and precautionary measures in the face of scientific uncertainties. These principles need to be considered in the context of risk management, which provides a framework and approach for setting the best course of action under uncertainty. Risk management is about balancing different concerns, and acknowledges the need for risk appetite and cautionary (precautionary) measures.

We refer to discussions in Chapters 18 and 19.

Final remarks

To quantitatively express uncertainties, an adequate representation is required, and probability is the natural choice as it meets some basic requirements for such a representation (Bedford and Cooke, 2001, p. 20), as mentioned in Chapter 18:

- **Axioms:** Specifying the formal properties of the uncertainty representation.
- **Interpretations:** Connecting the primitive terms in the axioms with observable phenomena.
- **Measurement procedures:** Providing, together with supplementary assumptions, practical methods for interpreting the axiom system.

Many types of uncertainty representation exist, but many fail when it comes to interpretation. We should not use a representation which has no clear interpretation. It is not sufficient to say that a measure expresses a degree of something. We need to know what it means that the measure is 0.2 instead of 0.4. Refer to the discussion in Cooke (2004), Flage *et al.* (2008) and Aven (2009b).

Bayesian analysis and, in particular, Bayesian updating represent a powerful tool for the systematic assessment of uncertainties. The key principles followed by this approach should be as follows (Aven, 2003): the focus is on quantities representing states of the world; these quantities are predicted in the risk analysis and probability is used as a measure of uncertainty related to the true values of these quantities. Probability models may be introduced to facilitate the assessments, but only if the models can be justified and their parameters given meaningful interpretations. Uncertainty assessments extending beyond the probabilistic analysis are, however, required as stressed repeatedly in this book.

Probabilistic risk analysis can also be used in the case of large uncertainties, when we have more or less ignorance about the consequences and outcomes of the activity studied. In this case the risk analysis can play a role in describing the uncertainties, but the main focus will be on the qualitative assessments of the uncertainty factors. Crude probabilities can be produced, but would not be given much weight as their basis is weak.

To analyse and treat risk we may benefit from crude categorization of the situations and problems considered. Examples of such categorizations, consistent with the above principles, are presented in Chapter 7; see also Aven and Renn (2009b).

References

Aven, T. (2003) *Foundations of Risk Analysis*, John Wiley & Sons, Ltd, Chichester.

Aven, T. (2007) A unified framework for risk and vulnerability analysis and management covering both safety and security. *Reliability Engineering and System Safety*, **92**, 745–754.

Aven, T. (2009) A new scientific framework for quantitative risk assessments. *International Journal of Business Continuity and Risk Management*, **1**, 67–77.

Aven, T. (2009b) On the interpretations of non-probabilistic uncertainty representations in a reliability and risk analysis context. Submitted for possible publication.

Aven, T. and Renn, O. (2009a) On risk defined as an event where the outcome is uncertain. *Journal of Risk Research*, **12**, 1–11.

Aven, T. and Renn, O. (2009b) The role of quantitative risk assessments for characterizing risk and uncertainty and delineating appropriate risk management options, with special emphasis on terrorism risk. *Risk Analysis*, **29**, 587–600.

Bedford, T. and Cooke, R. (2001) *Probabilistic Risk Analysis. Foundations and Methods*, Cambridge University Publishing Ltd, Cambridge.

Cooke, R. (2004) The anatomy of the squizzel. The role of operational definitions in representing uncertainty. *Reliability Engineering and System Safety*, **85**, 313–319.

Flage, R. and Aven, T. (2009) Expressing and communicating uncertainty in relation to quantitative risk analysis (QRA). *Reliability and Risk Analysis: Theory and Applications*, **2**, 9–18.

Flage, R., Aven, T. and Zio, E. (2009) Alternative representations of uncertainty in system reliability and risk analysis – review and discussion, in S. Martorell, C. Guedes Soares and J. Barnett (eds), *Safety, Reliability and Risk Analysis: Theory, Methods and Applications*, CRC Press, Boca Raton, FL.

Kaplan, S. and Garrick, B.J. (1981) On the quantitative definition of risk. *Risk Analysis*, **1**, 11–27.

Lindley, D.V. (2006) *Understanding Uncertainty*, John Wiley & Sons, Inc., Hoboken, NJ.

Paté-Cornell, M.E. (1996) Uncertainties in risk analysis: six levels of treatment. *Reliability Engineering and System Safety*, **54** (2–3), 95–111.

de Rocquigny, E., Devictor, N. and Tarantola, S. (eds) (2008) *Uncertainty in Industrial Practice. A Guide to Quantitative Uncertainty Management*, John Wiley & Sons, Ltd, Chichester.

Rosa, E.A. (1998) Metatheoretical foundations for post-normal risk. *Journal of Risk Research*, **1**, 15–44.

Singpurwalla, N. (2006) *Reliability and Risk. A Bayesian Perspective*, John Wiley & Sons, Ltd, Chichester.

Shrader-Frechette, K. (1985) *Risk Analysis and Scientific Method: Methodological and Ethical Problems with Evaluation Societal Hazards*, Reidel, Dordrecht.

Further reading

Aven, T. (2008) *Risk Analysis*, John Wiley & Sons, Ltd, Chichester.

Index

Misconceptions of Risk T. Aven